WORLDS, TIMES AND SELVES

WORLDS, TIMES AND SELVES

A. N. Prior

and

Kit Fine

UNIVERSITY OF
MASSACHUSETTS PRESS
AMHERST

First published in 1977 by
Gerald Duckworth & Company Limited
The Old Piano Factory
43 Gloucester Crescent, London NW1

Published in the United States of America by the
University of Massachusetts Press

International Standard Book Number 0–87023–227–4
Library of Congress Catalog Card Number 76–450421

Printed in Great Britain by
Ebenezer Baylis and Son Limited
The Trinity Press, Worcester, and London

CONTENTS

PREFACE

Before he died, Prior was working on a book to be entitled 'Worlds, Times and Selves'. This book was to deal, in one way or another, with the interplay between modal or tense logic, on the one hand, and quantification theory on the other. One of its main concerns was to show that modal and tense logic could stand on their own, that talk of possible worlds or instants was to be reduced to them rather than the other way round.

Unfortunately, only the first chapter was completed. There were jottings for other chapters, but they were far from complete. However, it is clear that some of Prior's recently published papers would have been incorporated into the book, though probably in considerably modified form. So what I have tried to do is to collate the published and unpublished material in such a way that the result is as close as possible to the book he had in mind.

This should explain the distribution of the unpublished material. The completed chapter appears, as it should, as the first paper of the collection. However, the other passages of unpublished material appear as supplements, in 3 and 7, to already published papers. This is because they are quite obviously expansions or elaborations of those papers. In order to avoid repetition, I have omitted some initial sections from the supplementary papers; and this accounts for their abrupt beginnings. I have also not used some other unpublished material, either because it was expository or because it was too fragmentary to be of interest.

The first paper explains in very simple terms the parallel between modal logic and quantification theory. It is a good introduction to the technical and philosophical problems that arise in the later papers.

The next three papers deal with the egocentric counterpart to ordinary tense or modal logic. They introduce the operator Q that picks out those propositions that correspond to instants, worlds or selves, as the case may be. The last sections of 2 and 4 and most of the supplement 3 are concerned with the formal development of Q or cognate notions.

The last three of Prior's papers, 5–7, deal with the problem of embedding the theory of instants or possible worlds within orthodox tense and modal logic respectively. Chapter 5 attempts to see how far the opposite view can be maintained. Chapter 6 is a particularly rich paper. It deals, among other things, with a world-calculus for the system Q, the logic of significance, and the extension of embedding results to possibilist quantifiers. The supplement elaborates further on some of these topics.

In his book, Prior would certainly have said more on this question of embedding. In the postscript, I have tried to fill this gap by discussing in detail his proposal for explaining instants and possible worlds within tense or modal logic. I had intended to write on his whole philosophy of time and modality; but, for reasons of space, I decided to stick to this more limited topic.

I should like to thank the editors of *Nous*, *L'Age de la Science*, *American Philosophical Quarterly*, and *Theoria* for permission to publish papers originally published by them. I should also like to thank Mary Prior, Anthony Kenny and Hans Kamp. They all, in their own ways, helped me to produce this collection. Tom Dimas and Mike Ferejohn prepared the indexes.

Kit Fine

*1. The Parallel between Modal Logic and Quantification Theory**

1. *General plan of the present work*

In what follows I shall be studying various inter-relations between two major branches of logic. One of these is modal logic, that is to say the study of the logical features of necessity and possibility, in which we come up with such laws as 'Whatever is necessary is the case' ('Whatever is bound to be so is so', 'If necessarily *p* then *p*') and 'Whatever is the case is possible' ('Whatever is so could be so', 'If *p* then possibly *p*'). The other major branch of logic we shall consider is the predicate calculus, or quantification theory, that is to say the study of the logical behaviour of 'Everything' and 'Something', in which we come up with such laws as 'What is true of everything is true of any given thing' ('What goes for everything goes for this', 'If for all *x*, *fx*, then *fa*') and 'What is true of a given thing is true of something' ('What goes for this goes for something', 'If *fa* then for some *x*, *fx*'). I shall also frequently cast a side glance at a minor branch of logic which is in some ways in between the other two, namely that fragment of the logic of temporal determination ('tense logic') in which we study the logical behaviour of 'Always' and 'At some time', in which we come up with such laws as 'What is always the case is the case now' ('If always *p* then *p*') and 'What is the case now is the case at some time' ('If *p* then *p* at some time').

It has for a long time been evident that there are close structural analogies between modal logic and quantification theory. The laws cited above, for example, are very similar. In each case we have a simple form (the plain *p*, or the plain *fa*) and a 'strong' prefix

* [The references in this piece are to the later chapters of the book that were never completed.]

('Necessarily' or 'For all *x*') yielding a form ('Necessarily *p*' or 'For all *x*, *fx*') which implies but is not implied by the simple form ('Necessarily *p*' implies but is not implied by the plain *p*, and 'For all *x*, *fx*' implies but is not implied by the plain *fa*) and a 'weak' prefix ('Possibly' or 'For some *x*') yielding a form ('Possibly *p*' or 'For some *x*, *fx*') which is implied by but does not imply the simple form ('Possibly *p*' is implied by but does not imply the plain *p*, and 'For some *x*, *fx*' is implied by but does not imply the plain *fa*). There are other parallels too, and later in this chapter I shall go into these in some detail, and systematically.

What is the philosophical significance of these parallels? A possibility which immediately suggests itself is that the modal expressions 'necessarily' and 'possibly' are disguised quantifications of some sort. This is already suggested by some common locutions, e.g. for 'Possibly *p*' we sometimes say 'There is some chance that *p*' or 'In some cases, *p*', and for 'Necessarily *p*' we sometimes say 'There is no chance that not *p*' or 'In every case, *p*'. And at least since Leibniz, modalities have been frequently thought of as quantifications over 'possible worlds'—it is necessary that *p* if *p* is true in all possible worlds, and possible if it is true in some. This view has been given great precision in modern accounts, due to Kripke and others, of the 'semantics' of modal systems, i.e. of the conditions of truth and validity of modal formulae. Tenses have been similarly presented as quantifications over the instants at which tensed propositions are true. These various attempted 'reductions' of modal logic and tense logic to quantification theory will be considered and criticised in Chapter 2.

It is much less usual to turn the parallels between modal logic and quantification in the opposite direction, and present quantification theory, or part of it, as being a disguised form of modal logic. Such a move, all the same, is in principle possible, and there is more to be said for it than one might at first imagine. Possibility and necessity are on the face of it rather 'metaphysical' notions, and in an intellectual climate which is rather hostile to metaphysics it is natural that there should be attempts to explain them away. Ordinary predicate logic or quantification theory, however, has its own metaphysical presuppositions, in particular the presupposition that the world consists of *things* of which this and that may be predicated, either individually, as when we say that *x* is *f*, or generally, as when we say that something or everything is *f*; and those who are uneasy about this presupposition might welcome the replacement of the machinery of ordinary predicate logic by something more like the machinery of modal logic. This possibility will be explored in Chapter 3, and taken somewhat further in Chapter 4. Chapter 4 is concerned with the development of modal and tense-logical analogues of what might be called 'individualising quantification', i.e. of the sort of quantification involved in saying that exactly one thing is *f*.

In Chapter 5 I shall turn back again to the formal parallels between modal logic and tense logic on the one hand and predicate logic or quantification theory on the other; I shall take a harder look at these parallels and see if they really extend as far as they at first appear to do. For example, just as 'Not everything is *f*' entails 'Something is not *f*', so 'Not necessarily *p*' seems to entail 'Possibly not *p*', and 'Not always *p*' seems to entail 'At some time not *p*'. But do these latter entailments really hold? I shall argue that they do not, and that, for example, 'This has not always been either green or not green' does not entail 'This has at some time been neither green nor not green', the former being true but the latter false if the thing referred to has not always existed. On similar grounds, we might deny that 'This did not have to be *f*' entails 'This could have failed to be *f*', i.e. this case of 'Not necessarily *p*' does not entail the corresponding case of 'Possibly not *p*'. I shall develop in detail a system of modal logic which allows for exceptions of this sort, and which does not seem to be interpretable as a disguised piece of quantification theory.

I shall not, however, at first press this conclusion too hard. For it may be that our predicate logic as well as our modal logic and tense-logic needs to be more sophisticated than it commonly is, and that if we revise that also, our parallels can be reinstated. In Chapter 5 I shall take this line of thought as far as I think it can be taken, and then indicate where I think it breaks down. To put the matter briefly, I think that even with these new sophistications the parallel breaks down when the modal or tense logic which we are trying to interpret as a disguised quantification theory is not just a modalised or tensed *propositional* calculus but a modalised or tensed *predicate* calculus, in which 'Necessarily' and 'Possibly', or 'Always' and 'At some time', are not merely compared or contrasted with 'Everything' and 'Something', but used in conjunction with these. That is, we consider at this point the area of logic in which we find such laws as the entailment by 'There is something that could be *f*' of 'It could be that something is *f*'. On the face of it, these could be read as propositions involving two quantifiers, one explicit and one disguised; we might read the premiss and conclusion respectively as 'There is something that has some chance of being *f*' and 'There is some chance that something is *f*'. So the logic of modalised quantification *might* be just a special application of the logic of multiple ordinary quantification. But just this suggestion, I shall argue, cannot be fully carried through, even when our underlying quantification theory is of quite a subtle kind.

All the same, there is a genuine theory of 'possible states of affairs' that may be developed within a sufficiently rich modal logic, and a genuine theory of 'instants' which may be developed within a sufficiently rich tense logic, and a sketch of these developments is given in Chapter 6. There is, indeed, the beginning of such a development in Chapter 4, on 'individualising quantifiers', but that development is in terms of the rather naive modal logic considered in Chapters 1 to 3; Chapter 6 employs the subtler modal system set up in Chapter 5.

2. *The parallel between modality and quantification*

The first systematic logical work to include both a discussion of quantification and a discussion of modality is Aristotle's *De Interpretatione*. What is to be 'interpreted' in this work is the pair of answers 'Yes' and 'No' which may be given to what was called a 'dialectical' question. So we divide propositions into contradictory pairs, each pair having an 'affirmative' member, expanding the answer 'Yes' to a given question, and a 'negative' member, expanding the answer 'No' to the same question. Saying 'Yes' to 'Is Socrates white?', for example, is tantamount to saying 'Socrates is white', and saying 'No' to it is tantamount to saying 'Socrates is not white'; these propositions form a contradictory pair. Here the only difference between the affirmation and the corresponding denial is that the latter has 'is not' where the former has 'is'. But it is not always as simple as that, and the first complication which Aristotle considers (ch. 7) is that which arises when the subject is implicitly or explicitly preceded by a sign of quantity. Though Aristotle does not put the matter quite like this, we might say that he is inviting us here to consider and distinguish between the two questions 'Is every man white?' and 'Is any man white?' 'Yes' and 'No' to the first mean 'Every man is white' and 'Not every man is white' respectively, and 'Yes' and 'No' to the second mean 'Some man is white' and 'No man is white' respectively. These two contradictory pairs Aristotle does explicitly distinguish, and he warns us against giving 'No man is white', for example, as the contradictory of 'Every man is white'. Both of these, he points out, may be false. 'No man is white', we might say, is not what we mean when we say 'No' to 'Is every man white?', but is the negative answer to a different question, namely 'Is any man white?', and there is nothing in logic that forbids us to answer 'Yes' to that question and 'No' to the other.

Then later, in ch. 12, he indicates that we must be similarly wary when considering not simply whether something is so, but whether it must be so or whether it may be so. 'No' to the latter does not just mean 'It may not be so' (*that* would expand 'No' to the former), but rather 'It cannot be so'. And here Aristotle sees that we can distinguish the two issues by talking, like Miss Anscombe, about an 'external' and an 'internal' negation. That something may not be so means that *its not being so is possible* ('internal' negation); that it cannot be so, means that *its being so is not possible* ('external' negation). The schoolmen said that in the former case we deny the *dictum*, i.e. the proposition about whose possibility we are inquiring, while in the latter case we deny the 'mode', in this case the possibility of the thing. Aristotle did not quite see, or at all events did not quite say, that we can deal with the signs of quantity similarly, but the schoolmen did have this matter nicely taped, in their theory of 'equipollence' (which, however, does owe something to Aristotle's discussion of negative subjects and predicates in ch. 10 of *De Interpretatione*).

What the schoolmen said was that putting 'not' before the sign of quantity, i.e. before the whole proposition, was equivalent to putting it before the predicate of the proposition with the opposite quantity, i.e. 'Not every man is white' is equivalent to 'Some man is not white', and 'Not (some man is white)', or as we would say 'Not any man is white', to 'Every man (is not white)'. They also brought the negative sign of quantity 'No' into it, observing that 'No man is white' is equivalent to 'Every man (is not white)' and so to 'Not (some man is white)' ('Not any man is white'), and 'Not (no man is white)' is equivalent to 'Some man is white', and this last therefore to 'Not every man is not white'. Further, 'Every man is white,' being equivalent to 'Not (some man is not white)' ('Not any man is not white'), is equivalent to 'No man (is not white)'; and 'Not every man is white', being equivalent to 'Some man is not white', is equivalent to 'Not (no man is not white)'. In brief,

(1) *Every* = *Not any not* = *No not*
(2) *Not every* = *Some not* = *Not no not*
(3) *Every not* = *Not any* = *No*
(4) *Not every not* = *Some* = *Not no.*

Then the schoolmen applied the same term *aequipollentia* to equivalences involving internal and external negations of modal propositions, in particular these:

(1) *Necessary that . . .*
= *Not possible that not . . .*
= *Impossible that not . . .*
(2) *Not necessary that . . .*
= *Possible that not . . .*
= *Not impossible that not . . .*
(3) *Necessary that not . . .*
= *Not possible that . . .*
= *Impossible that . . .*
(4) *Not necessary that not . . .*
= *Possible that . . .*
= *Not impossible that . . .*

Here the principal rule is that a 'not' *before* either 'necessary' or 'possible' (like one before 'every' or 'some') is equivalent to a 'not' *after* the other member of this pair.

Modality and quantification may be combined both 'internally' and 'externally' not only with negation but also with implication, conjunction and disjunction, and with these combinations also there is an exact parallelism. For example, both necessity and universal quantification 'distribute over implication'; that is, from 'Necessarily (if p then q)' we may infer 'If necessarily p then necessarily *q*' (a point noted by Aristotle in the *Prior Analytics*, book 1, ch. 15), just as from 'Anything, if it is *f*, is *g*' we may infer 'If everything is *f* then everything is *g*'. As to conjunction, it is important to distinguish, e.g., 'It is possible to write and possible not to write' from 'It is possible to write-and-not-write'; and analogously it is important to distinguish 'Someone is writing and someone is not' from 'Someone is both writing and not writing'. And as to disjunction, we have to distinguish 'Necessarily either *p* or not *p*' from 'Either necessarily-*p* or necessarily-not-*p*', just as we have to between 'Every animal is either rational or irrational' from 'Either every animal is rational or every animal is not rational' (an example from Peter of Spain's *Summulae Logicales*).

3. *The formalisation of modal logic*

Enough has been said, in a comparatively informal way, to make it clear that the analogies between modality and quantification are close and considerable. I shall now make these analogies more precise by setting up a symbolic calculus designed to represent the logical features of the necessary and the possible, and another designed to represent those of 'all' and 'some'; given these, I shall show how the symbolism originally devised to represent modal logic can be equally well regarded as representing an important part of quantification theory. For both purposes we begin with a representation of the classical truth-functional propositional calculus, a branch of logic which is presupposed in, or is a part of, both modal logic and quantification theory. For this I shall use the symbolism of Łukasiewicz, with the small letters *p*, *q*, *r* etc. for propositional variables, and the following symbolic complexes for truth functions:

$N\alpha$ for 'It is not the case that α'
$C\alpha\beta$ for 'If α then β'
$K\alpha\beta$ for 'Both α and β'
$A\alpha\beta$ for 'Either α or β'
$E\alpha\beta$ for 'If and only if α then β'.

In both the modal and the quantificational systems, it will be assumed that all tautologies of the classical two-valued propositional calculus are theorems.

For modal logic, we add to these the following two further complexes:

$L\alpha$ for 'Necessarily α'
$M\alpha$ for 'Possibly α',

with their own characteristic postulates. At this point, a variety of choices confront us. In the first place, there are modal systems of varying degrees of strength, differing in the laws that they contain. To begin with, we shall consider postulates for the system which Lewis called S5. But here again, we have a variety of choices, i.e. there are a number of different selections of postulates which are known to yield precisely the theorems of Lewis's S5. Here, I shall give four. In the first, *L* ('Necessarily') is taken as undefined, and *M* ('Possibly') is defined as *NLN* ('Possibly α' as 'Not necessarily not α'), and we subjoin to the tautologies of propositional calculus the axiom-schema

A1. $CL\alpha\alpha$

(or alternatively, the axiom *CLpp* with a rule of substitution for propositional variables). The equivalence of *M* to *NLN* is one of the old rules of modal 'equipollence' mentioned in the last section, and the axiom-schema, 'If necessarily α then α', embodies the simple principle mentioned in Section 1, that what is necessary is the case, or as the schoolmen put it, *A necesse esse ad esse valent consequentia* ('It is justifiable to infer "is" from "necessarily is"'). For rules of inference we have

Detachment: If $\vdash\alpha$ and $\vdash C\alpha\beta$ we may infer $\vdash\beta$; and
RL: If $\vdash C\alpha\beta$ then $\vdash C\alpha L\beta$, provided that every variable in α falls within the scope of an *L*.

One important derived rule in this system is

RCL: If $\vdash C\alpha\beta$ then $\vdash CL\alpha L\beta$,

which we derive thus:

1. $C\alpha\beta$ (hypothesis)
2. $CC\alpha\beta CCL\alpha\alpha CL\alpha\beta$ (tautology; case of $CC\alpha\beta CC\gamma\alpha C\gamma\beta$)
3. $CL\alpha\beta$ (2, 1, A1, two detachments)
4. $CL\alpha L\beta$ (3, RL).

Another important derived rule is

Necessitation: If $\vdash\alpha$ then $\vdash L\alpha$.

We derived this rule as follows

1. α (hypothesis)
2. $C\alpha CCLpLp\alpha$ (tautology, being of the form $C\alpha C\beta\alpha$)
3. $CCLpLp\alpha$ (1, 2, detachment)
4. $CCLpLpL\alpha$ (3, RL)
5. $CLpLp$ (tautology, being of the form $C\alpha\alpha$)
6. $L\alpha$ (4, 5, detachment).

Theorem-schemata which are provable in this system include the following:

T1. $CLC\alpha\beta C\alpha\beta$ (case of A1)
T2. $CCLC\alpha\beta C\alpha\beta CCL\alpha\alpha CLC\alpha\beta CL\alpha\beta$ (tautology; case of $CC\alpha C\beta\gamma CC\delta\alpha C\alpha C\delta\gamma$)
T3. $CLC\alpha\beta CL\alpha\beta$ (T2, T1, A1, two detachments)
T4. $CCLC\alpha\beta CL\alpha\beta CKLC\alpha\beta L\alpha\beta$ (tautology; case of $CC\alpha C\beta\gamma CK\alpha\beta\gamma$)
T5. $CKLC\alpha\beta L\alpha\beta$ (T4, T3, detachment)
T6. $CKLC\alpha\beta L\alpha L\beta$ (T5, RL)
T7. $CCKLC\alpha\beta L\alpha L\beta CLC\alpha\beta CL\alpha L\beta$ (tautology; case of $CCK\alpha\beta\gamma C\alpha C\beta\gamma$)
T8. $CLC\alpha\beta CL\alpha L\beta$ (T7, T6, detachment).

(This last embodies the Aristotelian principle that where β necessarily follows from α, the necessity of β follows from that of α.) Where every variable in α falls within the scope of an L, we have further:

T9. $C\alpha L\alpha$ ($C\alpha\alpha$, RL)
T10. $CCLC\alpha\beta CL\alpha L\beta CC\alpha L\alpha CLC\alpha\beta C\alpha L\beta$ (tautology; case of $CC\alpha C\beta\gamma CC\delta\beta C\alpha C\delta\gamma$)
T11. $CLC\alpha\beta C\alpha L\beta$ (T10, T8, T9, two detachments)
T12. $CNL\alpha LNL\alpha$ (case of T9).

To obtain a second postulate-set for S5, we may replace RL by Necessitation, T8 and T9 (which, remember, has the proviso that every variable in α must fall within the scope of an *L*), or by Necessitation, T8 and T12 (the basis used with A1 and Detachment by Gödel). And to obtain a third, we may replace RL by Necessitation and T11 (which, again, has the proviso that every variable in α must fall within the scope of an *L*). The derivation of our original postulates from these new ones need not be given here; all that is needed is to adapt Lemmon's derivation of a similar one from Lewis's original postulates for S5.

Further theorem-schemata derivable from our original postulates are these:

T13. $CLN\alpha N\alpha$ (case of A1)
T14. $CCLN\alpha N\alpha C\alpha NLN\alpha$ (tautology; case of $CC\alpha N\beta C\beta N\alpha$)
T15. $C\alpha NLN\alpha$ (T14, T13, detachment)
T16. $C\alpha M\alpha$ (T15, Df. *M*).

(This last is the principle, mentioned in Section 1, that what is so could be so, or as the schoolmen put it, *Ab esse ad posse valet consequentia*, 'It is justifiable to infer "can be" from "is" '.) And we have the derived rule

RM: If $\vdash C\alpha\beta$ then $\vdash CM\alpha\beta$, provided that every variable in β falls within the scope of an *L*.

This last condition we may describe as β's being 'fully modalised'. Given β's full modalisation, we derive RM thus:

1. $C\alpha\beta$ (hypothesis)
2. $CC\alpha\beta CN\beta N\alpha$ (tautology)
3. $CN\beta N\alpha$ (2, 1, detachment)
4. $CN\beta LN\alpha$ (3, RL)

5. $CCN\beta LN\alpha CNLN\alpha\beta$ (tautology; case of $CCN\alpha\beta CN\beta\alpha$)
6. $CNLN\alpha\beta$ (5, 4, detachment)
7. $CM\alpha\beta$ (6, Df. *M*).

It may be noted that β will be fully modalised if each of its variables falls within the scope of an *M*, since that is to fall within the scope of an *NLN*, and so within the scope of an *L*. We also have in our system, for any α (fully modalised or not), the theorem-schemata

T16. $CCL\alpha\alpha CL\alpha NN\alpha$ (tautology; case of $CC\alpha\beta C\alpha NN\beta$)
T17. $CL\alpha NN\alpha$ (T16, A1, detachment)
T18. $CL\alpha LNN\alpha$ (T17, RL)
T19. $CCL\alpha LNN\alpha CL\alpha NNLNN\alpha$ (tautology; case of $CC\alpha\beta C\alpha NN\beta$)
T20. $CL\alpha NNLNN\alpha$ (T19, T18)
T21. $CL\alpha NMN\alpha$ (T20, Df. *M*)
T22. $CNMN\alpha L\alpha$ (similar proof, using $CC\alpha\beta CNN\alpha\beta$).

We may also prove, for any function *f* symbolised in the system, that if $\vdash C\alpha\beta$ and $\vdash C\beta\alpha$ then $\vdash Cf(\alpha)f(\beta)$ and $\vdash Cf(\alpha)f(\beta)$, by $C\alpha\alpha$, $CC\alpha\beta CN\beta N\alpha$, $CC\alpha\beta CC\beta\gamma C\alpha\gamma$, $CC\alpha\beta CC\gamma\alpha C\gamma\beta$, RCL, and induction on the complexity of *f*. (In citing these particular formulae I am assuming that our undefined truth-functions are $N\alpha$ and $C\alpha\beta$, the rest abridging complexes of these. If we used, say *N* and *K* as primitive instead of *N* and *C*, $CC\alpha\beta CC\beta\gamma C\alpha\gamma$ and $CC\alpha\beta CC\gamma\alpha C\gamma\beta$ would have to be replaced by $CC\alpha\beta CK\alpha\gamma K\beta\gamma$ and $CC\alpha\beta CK\gamma\alpha K\gamma\beta$). Given this rule, T21 and T22 will enable us to replace *NMN* by *L*, or *L* by *NMN*, in any formula, exactly as we could if we had taken *M* as our undefined modal operator and defined *L* as *NMN*.

This last is precisely what we do in our fourth and final postulate set for S5. For this our one axiom-schema, instead of A1, is T16. $C\alpha M\alpha$, and our rules are Detachment and the above RM, i.e. from $\vdash C\alpha\beta$ to infer $\vdash CM\alpha\beta$, provided that β is fully modalised. We now define 'full modalisation', however, as having every variable falling within the scope of an *M*. Having every variable falling within the scope of an *L* will of course be a case of 'full modalisation' in this new sense, since every variable will then fall in the scope of an *NMN*, and so of an *M*. An important derived rule of this system is

RCM: If $\vdash C\alpha\beta$, then $\vdash CM\alpha M\beta$,

which we derived as follows:

1. $C\alpha\beta$ (hypothesis)
2. $C\beta M\beta$ (case of T16)
3. $CC\alpha\beta CC\beta M\beta C\alpha M\beta$ (tautology; case of $CC\alpha\beta CC\beta\gamma C\alpha\gamma$)
4. $C\alpha M\beta$ (3, 1, 2, two detachments)
5. $CM\alpha M\beta$ (4, RM).

Theorem schemata in this system include

T23. $CN\alpha MN\alpha$ (case of T16)
T24. $CCN\alpha MN\alpha CNMN\alpha\alpha$ (tautology; case of $CCN\alpha\beta CN\beta\alpha$)
T25. $CNMN\alpha\alpha$ (T24, T23, detachment)
A1. $CL\alpha\alpha$ (T25, Df. L)
T26. $CC\alpha M\alpha CNN\alpha M\alpha$ (tautology; case of $CC\alpha\beta CNN\alpha\beta$)
T27. $CMN\alpha M\alpha$ (T26, T16, detachment)
T28. $CMNN\alpha M\alpha$ (T27, RM)
T29. $CCMNN\alpha M\alpha CNNMNN\alpha M\alpha$ (tautology; case of $CC\alpha\beta CNN\alpha\beta$)
T30. $CNNMNN\alpha M\alpha$ (T29, T28, detachment)
T31. $CNLN\alpha M\alpha$ (T30, Df. L)
T32. $CM\alpha NLN\alpha$ (similar proof, using $CC\alpha\beta C\alpha NN\beta$).

Using $C\alpha\alpha$, $CC\alpha\beta CN\beta N\alpha$, $CC\alpha\beta CC\beta\gamma C\alpha\gamma$, $CC\alpha\beta CC\gamma\alpha C\gamma\beta$ and RCM, we may prove inductively that if $\vdash C\alpha\beta$ and $\vdash C\beta\alpha$ then $\vdash Cf(\alpha)f(\beta)$ and $\vdash Cf(\beta)f(\alpha)$, for any f constructible in the system. Hence T31 and T32 enable us to replace M by NLN, or NLN by M, anywhere in a formulae, as we can in the other systems in virtue of M's definition. A1 is proved above, and the rule RL may be derived in the new system thus, assuming α to be fully modalised:

1. $C\alpha\beta$
2. $CC\alpha\beta CN\beta N\alpha$ (tautology)
3. $CN\beta N\alpha$ (2, 1, detachment)
4. $CMN\beta N\alpha$ (4, RM)
5. $CCMN\beta N\alpha C\alpha NMN\beta$ (tautology; case of $CC\alpha N\beta C\beta N\alpha$)

6. $C\alpha NMN\beta$ (5, 4, detachment)
7. $C\alpha L\beta$ (6, Df. L).

So our first and last systems are now proved equivalent.

4. *The formalisation of first-order predicate logic*

I now construct a symbolic calculus designed to represent at least some of the logical features of 'all' and 'some'. For this, two new sorts of variables are introduced, the individual name variables *x*, *y*, *z*, etc. and the predicate variables *f*, *g*, *h*, etc. Some of these latter stand for 'one-place' predicates, which construct a well-formed formula, say *fx*, when attached to a name-variable; others stand for 'two-place' predicates, which construct a well-formed formula, say *gxy*, when attached to two name-variables; others for 'three-place' predicates, and so on. Then there are the two quantifiers Π, 'For all', and Σ, 'For some', which form well-formed formulae when attached to a variable followed by a well-formed formula, as in Π*xfx*, 'For all *x*, *fx*', and Π*x*Σ*ygxy*, 'For all *x*, for some *y*, *gxy*'. In these examples the variables 'bound' by the quantifiers are name-variables, and we shall confine ourselves in the present chapter to that portion of quantification theory in which these are the only variables that immediately follow quantifiers; that is, we confine ourselves at present to the 'first-order' predicate calculus.

The above, at least, is what might be called the standard Łukasiewicz symbolism, but of course there is nothing sacred about its details, and in the present section they will be modified as follows: First, we shall use *p*, *q*, *r*, etc. for predicate variables as well as for propositional variables, regarded a proposition as a 'no-place' predicate, i.e. as forming a well-formed formula when attached to *no* name-variables. Secondly, we shall use *L* and *M* instead of Π and Σ respectively. As we are not symbolising modal logic in this section, this will not result in any ambiguity. So now we write *Lxpx* for 'For all *x*, *px*' and *LxMyqxy* for 'For all *x*, for some *y*, *qxy*'.

Using *v* as a metalogical symbol for any name-variable, and α, β, etc. before as metalogical symbols for well-formed formulae, we lay down the following axiom-schema:

A1v: $CLv\alpha\beta$, where β is either the same as α or differs from it only in the replacement of all occurrences of *v* which are free in α by some other name-variable for which *v* is free in α.

(That is, β may have all occurrences of *v* in α which are not within the scope of a further *Lv* forming part of α itself, replaced by some other name-variable *w*, provided that no such 'free' occurrence of *v* in α falls within the scope of an *Lw*.) Cases of A1 would be *CLxpxpx* and *CLxpxpy*. *CLxMxqxyMxqyy*, *CLxMxqxyMyqyy* and *CLxMyqxyMyqyy* would all be excluded by the provisos about freedom (and would be false, or capable of falsehood, if *qxy* means '*x* is other than *y*', so that, e.g. the third formula meant that if everything has something other than it, then something is other than itself). And to detachment we add the rule

RLv: If ⊢*C*αβ then ⊢*C*α*Lv*β, if the variable *v* does not occur free in α.

The universal quantifier *L* is taken as undefined, and *Mv* defined as *MLvN*.

It is fairly obvious that all the derivations in the preceding section from our first basis for modal logic could be matched by analogous derivations in first-order predicate logic, the only difference being in the greater variety of the well-formed formulae for which α, β, etc. may stand, and the appearance of a name-variable after each occurrence of *L* or *M*. For example, our proof of T16, *C*α*M*α, can be matched by the following:

T13v. *CLvN*α*N*α (case of A1)
T14v. *CCLvN*α*N*α*C*α*NLvN*α (tautology; case of *CC*α*N*β*C*β*N*α)
T15v. *C*α*NLvN*α (T14v, T13v, detachment)
T16v. *C*α*Mv*α (T15v, Df. *M*).

Alternatively, we could take the 'particular' quantifier *M* as undefined, and define *Lv* as *NMvN*, replace A1v by T16v, and replace RLv by

RMv: If ⊢*C*αβ, then ⊢*CMv*αβ, if *v* does not occur free in β.

This basis could be shown equivalent to our first one by steps exactly analogous to those by which we showed our fourth basis for modal logic equivalent to our first. This is just a case of the general point that any derivation in our modal system can be matched by a derivation in our predicate logic, with *L* and *M* replaced by *Lv* and *Mv* throughout.

Is it also true that any derivation in our predicate logic can be matched by one in our modal logic with the name-variables deleted? Not quite; for consider the following (in which I shorten the presentation by not giving in full the relevant cases of the tautologies cited):

1. *CKpxNpyMyKpxNpy* (case of new T16v)
2. *CMyKpxNpyMxMyKpxNpy* (case of new T16v)
3. *CKpxNpyMxMyKpxNpy* (1, 2, *CCαβCCβγCαγ*)
4. *CpxCNpyMxMyKpxNpy* (3, *CCKαβγCαCβγ*)
5. *CMxpxCNpyMxMyKpxNpy* (4, RMv)
6. *CNpyCMxpxMxMyKpxNpy* (5, *CCαCβγCβCαγ*)
7. *CMyNpyCMxpxMxMyKpxNpy* (6, RMv).

In words, this tells us that if something is not *p* ('For some *y*, not *py*') and something is *p* ('For some *x*, *px*') then there are things of which one is *p* and another is *not* ('For some *x*, for some *y*, *px* and not *py*'). Intuitively, this seems quite acceptable. But now try dropping the name-variables, thus:

1. *CKpNpMKpNp* (case of T16)
2. *CMKpNpNMKpNp* (case of T16)
3. *CKpNpMMKpNp* (1, 2, *CCαβCCβγCαγ*)
4. *CpCNpMMKpNp* (3, *CCKαβγCαCβγ*)
5. *CMpCNpMMKpNp* (4, RM)
6. *CNpCMpMMKpNp* (5, *CCαCβγCβCαγ*)
7. *CMNpCMpMMKpNp* (6, RM).

The conclusion now states that if it could be that not *p* and could be that *p*, then it could be that it could be that both *p* and not *p*. Since 'could be that it could be' in S5 entails the plain 'could be' (from the tautology *CMpMp* we get *CMMpMp* by RM), this entails

8. *CMNpCMpMKpNp*

'If *p* could be false and could be true, then it could be both true and false', a rather famous fallacy in modal logic (attacked by Aristotle). In fact, however, the application of RM to obtain 5 in the derivation just given is not legitimate, since the consequent of 4, *CNpNMKpNp*, is not fully modalised (the first *p* does not fall within the scope of a modal operator). In the predicate-calculus derivation, however, we can put *Mx* in front of the antecedent of 4, because although its consequent *CNpyMxMyKpxNpy* contains free *y* it does not contain any free *x*, and it is a quantifier containing *x*, not one containing *y*, that we are introducing at this stage. The use of name-variables thus introduces moves that cannot be paralleled in modal logic, however similar the laws of modal operators and quantifiers might be. The basic reason is of course that different quantifiers in a formula may bind different name-variables. If, however, we remove this variety, a much closer analogy can be drawn.

Let us consider, then, not the first-order predicate calculus in its entirety, but a certain fragment of it, namely that fragment in which (a) we use one name-variable only, say *x*, and (b) our predicate-variables *p*, *q*, *r*, etc. are for one-place predicates only. So *Nxpx*, *LxCpxqx* and *MxLxpx* are formulae of this system, but *Cqxp*, *Cqxpxy* and *MxLypxy* are not. Our postulates we leave as before, but we apply them only to formulae of this restricted class. In this restricted application, the postulates can be somewhat simplified. A1v becomes simply

A1*x*: *CLx*αα,

since there is now no question of replacing *x* in α by some other variable. And RLv becomes

RLx: If ⊢*C*αβ then ⊢*C*α*Lx*β, provided that every predicate variable in α falls within the scope of an *Lx*,

since α can now only fail to contain *x* free by containing it bound, i.e. by each of its *px*'s etc. falling within the scope of some *Lx*.

We may call this system the uniform monadic first-order predicate calculus (monadic because all its predicates are one-place, uniform because only one name-variable is used), or UM1PC for short. Its one name-variable, we may note, may occur in a formula in two ways—bound or free. Bound, it functions in the usual way; *Lxpx* means 'Everything is *p*' and *Mxpx*, 'Something is *p*'. Free, it stands for some individual's name, the same throughout the formula. We might therefore regard the calculus as a language in which we have a name for one individual only, all other individuals being referred to only as covered by the general 'Something' and 'Everything' (or, a little more specifically, by 'Something that is *p*' or 'Everything that is *p*'). If this privileged individual be called 'Arthur', we may now read *CLxpxpx* (a case of A1) as 'If everything is *p*, then Arthur is *p*', and the more complicated theorem

CLxCMxpxqxCpxLxqx

as 'If it is true of everything that if something is *p* then that thing is *q*, then if Arthur is *p*, everything is *q*'.

In this system every *L*, every *M* and every predicate-variable is followed by an *x*, and these are the only places where any *x* appears. Its formulae would therefore be perfectly intelligible if we left all the *x*'s out, since we would know exactly where to put them back, in our minds, if we wanted to. But with all *x*'s deleted the formulae would become simply those of the modal calculus constructed in the last section. Moreover, the theorems with *x*'s deleted would be simply the theorems of S5. For (a) it is obvious that A1x or T16x would become A1 or T16, and (b) the proviso on the RLx or RMx would now operate in exactly the same cases as the proviso on RL or RM. We can, therefore, regard the modal system—considered purely as a symbolic calculus—as nothing more nor less than an abridged version of UM1PC. Conversely we could regard UM1PC, not as a fragment of the full predicate calculus, but as the modal system S5 with a superfluous symbol *x* systematically inserted into certain places in its formulae.

5. *Modalities and predicative complications*

Another way of reaching the same conclusions is the following: In most presentations of predicate calculus, there are no special devices for forming complex predicates as opposed to complex propositions. To symbolise '*x* is both-*p*-and-*q*', for instance, one just says 'Both *x*-is-*p* and *x*-is-*q*', *Kpxqx*. There is no reason in principle, however, why one should not introduce a special form for the predicate 'both-*p*-and-*q*', and define '*x* is both-*p*-and-*q*' as '*x* is *p* and *x* is *q*', i.e. define the predicative complications in terms of the propositional ones. So let us use the Łukasiewicz truth-functional symbols for complex predicates, and symbols of some other sort for complex propositions, say

'$-\alpha$' for 'Not α'
'$\alpha \rightarrow \beta$' for 'If α then β'
'$\alpha \,\&\, \beta$' for 'α and β'
'$\alpha \vee \beta$' for 'α or β'
'$\alpha \leftrightarrow \beta$' for '$\alpha$ if and only if β'.

We then define

'$(N\varphi)x$' as '$-(\varphi x)$'
'$(C\varphi\psi)x$' as '$\varphi x \rightarrow \psi x$'
'$(K\varphi\psi)x$' as '$\varphi x \,\&\, \psi x$'
'$(A\varphi\psi)x$' as '$\varphi x \vee \psi x$'
'$(E\varphi\psi)x$' as '$\varphi x \leftrightarrow \psi x$'.

If, at the propositional level, we used (say) '—' and '&' as undefined, and defined the rest in terms of these, then we could define '$(N\varphi)x$' as '$-(\varphi x)$' and '$(K\varphi\psi)x$' as '$\varphi x \,\&\, \psi x$', and thereafter define C, A and E in terms of N and K, getting the above relations of C, A, and E to the corresponding propositional symbols consequentially. For example, if we define $C\varphi\psi$ as $NK\varphi N\psi$ and $\alpha \rightarrow \beta$ as $-(\alpha \,\&\, \beta)$, we have

$$\begin{aligned}(C\varphi\psi)x &= (NK\varphi N\psi)x, \text{ by Df. } C\\ &= -(K\varphi N\psi)x, \text{ by Df. } N\\ &= -(\varphi x \,\&\, (N\psi)x), \text{ by Df. } K\\ &= -(\varphi x \,\&\, -\psi x), \text{ by Df. } N\\ &= \varphi x \rightarrow \psi x, \text{ by Df. } \rightarrow.\end{aligned}$$

In UM1PC it is also possible to regard quantifications, for which we may now use $\forall x$ ('For all *x*') and $\exists x$ ('For some *x*'), as forming complex predicates of our one individual, 'Everything is *p*' being equivalent to 'Arthur has the property of being *p* along with everything else' and 'Something is p' to 'Arthur has the property of being p if nothing else is'. We may use the form Lp for the former property and Mp for the latter; that is, we define

'$(L\varphi)x$' as '$\forall x\varphi x$'
'$(M\varphi)x$' as '$\exists x\varphi x$'.

Alternatively we define L as above, M as NLN, and $\exists x$ as $-\forall x-$ giving us

$$\begin{aligned}(M\varphi)x &= NLN\varphi)x, \text{ by Df. } M\\ &= -(LN\varphi)x, \text{ by Df. } N\\ &= -\forall x(N\varphi)x, \text{ by Df. } L\\ &= -\forall x-\varphi x, \text{ by Df. } N\\ &= \exists x\varphi x, \text{ by Df. } \exists.\end{aligned}$$

Or we define M as above, L as NMN and $\forall x$ as $-\exists x-$, equating $(L\varphi)x$ with $\forall x\varphi x$ indirectly.

Given these definitions, it is clear that any formula of S5, attached to the variable x, can be expanded to some formula, without the special predicative complications, of UM1PC. Conversely, any formula of UM1PC can be contrasted by the definitions to a single complex predicate attached to x, interpretable as the ascription of a single complex predicate to our individual Arthur. Thus $\forall xpx \rightarrow px$, 'If everything is p then Arthur is p', contracts to $(CLpp)x$, 'Arthur is p-if-he-is-p-along-with-everything-else', or 'Arthur is p-if-(p-along-with-everything-else)'. It is clear, too, that a formula of UM1PC will be a theorem of that calculus if and only if the predicate of x to which it contracts has the form of a theorem of S5. This is a corollary of the result of the last section, since if the x at the end of a predication like $(CLpp)x$ is taken from there and spread through the formula after each L, M or predicate-variable we just have a UM1PC formula in the notation of the last section.

2. *Egocentric Logic*

1. *Two ways of locating occurrences in times*

Tensed sentences such as 'Brown is ill', 'Brown has been ill' and 'Brown will be ill', have the peculiarity of being true at some times and false at others. This has led some logicians to say that such sentences do not explain genuine propositions but merely *predicates*, which can be thought of as *characterising* the times or instants 'at' which they could ordinarily be said to be true. We can at all events embed all propositions or quasi-propositions of a tensed language within a first-order theory of the earlier-later relation between instants, in ways which may be illustrated as follows:

(1) That *Brown is ill* is the case at the instant *a*
= (2) *a* is a Brown-being-ill-ish instant

(3) That *Brown has been ill* is the case at the instant *a*
= (4) *a* is a Brown-having-been-ill-ish instant
= (5) *a* is later than some Brown-being-ill-ish instant
= (6) Some instant earlier than *a* is a Brown-being-ill-ish one.

When stating these equivalences we employ an exterior untensed language which would be formalised as a first-order theory of the earlier-later relation between instants, and an (italicised) interior revised language which might nowadays be formalised by means of a quasi-modal logic with prefixes such as 'It has been the case that' functioning in the way that 'It is possible that' does in modal logic proper. In passing from (1) to (2), or from (3) to (4), we replaced a tensed sentence ('Brown is ill' or 'Brown has been ill') by a predicate of instants ('Brown-being-ill-ish' or 'Brown-having-been-ill-ish'); in passing from (4) to (5) we define the predicate 'Brown-having-been-ill-ish' in terms of the simple 'Brown-being-ill-ish' and the earlier-later relation, but still retain a wording which represents what we are reporting as the characterisation of the instant *a*; and in moving from (5) to (6) we show that what has been thus represented is 'really' a characterisation of a different instant.

(7) That *Brown will be ill* is the case at the instant *a*

can obviously be subjected to a similar series of transformations.

In our exterior tenseless language we assume that we have some way of referring to individual instants. In the interior tensed language no such instants are explicitly mentioned, but when we say the plain 'Brown is ill' it is understood that what is being implicitly characterised as Brown-being-ill-ish is the (unmentioned) *time of utterance*, and when we say 'Brown has been ill' or 'It has been the case that Brown is ill' it is understood that it is again the time of utterance that is being characterised as a Brown-having-been-ill-ish one, and that to represent the time of utterance as being thus characterised is a way of saying that some time earlier than that one is plain Brown-being-ill-ish.

Consider now a more complex case:

(8) That *Brown will have been ill* is the case at the instant *a*
= (9) *a* is a Brown-going-to-have-been-ill-ish instant
= (10) *a* is earlier than some Brown-having-been-ill-ish instant
= (11) Some instant later than *a* is a Brown-having-been-ill-ish one
= (12) Some instant later than *a* is later than some Brown-being-ill-ish one
= (13) Some instant later than *a* has some earlier instant which is Brown-being-ill-ish.

We may note also that

(11) = (14) Some instant later than *a* is one at which *Brown has been ill* is the case; and
(13) = (15) Some instant later than *a* has some earlier at which *Brown is ill* is the case.

It is clear that although the interior tensed language *mentions* no instants there is a sense in which it *implicitly refers* to the time of utterance, and by tensing what is implicitly said of the time of utterance it can indirectly characterise other times also, though these are referred to rather indefinitely.

If tenses are formed by attaching prefixes like 'It has been the case that' to the present tense, or to a complex with a present tense 'kernel', it is not always true to say that what is in the present tense is understood as a characterisation of the time of utterance; rather, it characterises whatever time we are taken to by the series of prefixes. The *presentness* of an event, we may say, is simply the *occurrence* of the event, and that is simply the event itself. But every complete tensed sentence characterises the time of utterance in some way or other, and other times only through their relation to that one.

2. *Two ways of locating properties in individuals*

If I say, not 'Brown is ill' but 'I am ill', the truth of this depends not only on when it is said but on who says it. It has been suggested, e.g. by Donald Davidson,[1] that just as the former dependence has not prevented the development of a systematic logic of tenses, so the latter should not prevent the development of a systematic logic of personal pronouns. But the machinery of tenses is in some ways very different from that of personal pronouns, and to bring out both the similarities and the differences I shall invent a stylised logical language, which I shall call Egocentric, in which properties *are* located in individuals in the same way as events are located in times by means of tenses.

In at least the most elementary tensed languages, as we have seen, instants or times are not mentioned, but tensed propositions are understood as directly or indirectly characterising the *un*-mentioned time of utterance. So in Egocentric, individuals must not be directly mentioned, but the propositions of Egocentric will be understood as directly or indirectly characterising the speaker. The propositions of Egocentric will in fact be subject-less predicates of common speech, and I shall represent them by participles, i.e. things like 'Standing', 'Sitting', 'Drinking' will be sentences of Egocentric, and on their own will mean what is ordinarily meant by 'I am standing', 'I am sitting', etc. We do sometimes use a

[1] Donald Davidson, 'Truth and Meaning', *Synthese*, XVII (September 1967).

language like this in subordinate clauses, as when we say 'I remember *being* at the meeting' (or 'I remember *having been* at the meeting') for 'I remember *that I was* at the meeting'.[1] So, formally, we have

(16) *Standing* is the case with *a*
= (17) 'I am standing' is true when said by *a*
= (18) *a* is standing.

To obtain oblique predications analogous to past and future tenses, we must find some relation between individuals which we can exploit in the way that tensed languages exploit the earlier-later relation between instants. Any relation which holds directly or indirectly between all individuals will do, but for simplicity's sake I shall assume with Leibniz that all individuals are arranged in a scale of comparative perfection, and use the form 'Inferior-to-standing' or 'Someone-more-perfect standing' to mean that I am less perfect than someone who is standing, i.e. that someone more perfect than me is standing. That is, we have the equivalences

(19) *Inferior-to-standing* (or *Someone-more-perfect standing*) is the case with *a*
= (20) 'I am inferior to someone standing' (or 'Someone more perfect than me is standing') is true when said by *a*
= (21) *a* is inferior to someone standing
= (22) Someone more perfect than *a* is standing.

The form 'Superior-to-standing' or 'Someone-less-perfect standing' is similarly understood. Here we have an exterior language, without the modalities 'Inferior-to-' and 'Superior-to-', in which individuals are mentioned in the normal way, which could be formalised as a first-order theory of the relation of comparative perfection between individuals; and an interior language (Egocentric) in which reference to individuals is achieved by those modalities which take us to and from the speaker. In passing from (19) to (21) the Egocentric *proposition* 'Inferior-to-standing' is replaced by the *predicate* 'inferior to someone standing', and in passing from (21) to (22) we see that what (21) represents as a characterisation of the speaker is

[1] My attention was drawn to this suggestive way of talking by a lecture by Miss J. Rountree.

what might be more straightforwardly represented as a simpler characterisation of someone else.

In a more complicated case, we have

(23) *Inferior to superior to standing* (or *Someone-more-perfect someone-less-perfect standing*) is the case with *a*
= (24) 'I am inferior to someone superior to someone standing' (or 'Someone more perfect than me has someone less perfect than him standing') is true when said by *a*
= (25) *a* is inferior to someone superior to someone standing
= (26) Someone more perfect than *a* (say *b*) has someone less perfect than him (i.e. than *b*) who is standing.

We may note also that

(25) = (27) *a* is inferior to someone with whom *Superior-to-standing* (or *Someone-less-perfect standing*) is the case
= (28) *a* is inferior to someone who can say truly 'Someone less perfect than me is standing'; and
(26) = (29) Someone more perfect than *a* (say *b*) has someone less perfect than him (i.e. than *b*) with whom *Standing* is the case
= (30) Someone less perfect than *a* (say *b*) has someone less perfect than him who can say truly 'I am standing'.

It is not true to say that such a form as 'Standing' is always understood as characterising the speaker; rather, it characterises whatever individual we are taken to by the series of prefixes. But every Egocentric sentence characterises the speaker in some way or other, and characterises other individuals only through their relation to the speaker.

3. *Derivation of other modes of temporal reference within tense logic*

Could Egocentric be an *adequate* language for talking about individuals? On the face of it not, since it has no devices for referring specifically to other individuals than the speaker. But could it be so enlarged, without losing its egocentric character, as to contain such devices?

There is a similar problem about tense logic; could that be so enlarged, without losing its tensed character, as to contain devices for referring to specific times? The answer is that it can, though what devices are available to us depends to some extent on the nature of the time-series. Quite simple devices are available if the series of instants is linear and if each instant has something which is true at that instant only; so to make our illustrations simple we shall adopt these assumptions.[1] We begin with some more equivalences:

(31) That p, is the case at the present only
= (32) *(It is the case that) p, but it has not been the case that p and will not be the case that p.*

(I italicise tensed propositions.) And

(33) That p, is the case at one instant only
= (34) At some time (*p, and it has not been the case that p, and it will not be the case that p*)
= (35) At some time, (32)
= (36) *Either* (32) *or it has been the case that* (32) *or it will be the the case that* (32).

A proposition of which (36) is true will serve to *identify* a particular instant in tense logic; we may say that an instant *is* such a proposition. Philosophically the most interesting proposition which is true at a given instant only is the conjunction of all the propositions which are then true, but for formal purposes any proposition true at that instant only will do as its tense-logical 'representative'. In what follows, I use A for the 'representative' of a, B for that of b, etc.

(37) That *Brown is ill* is the case at a
= (38) At some time, (*it is the case that*) *both A and Brown is ill*[2]
= (39) *Either* (*both A and Brown is ill*) *or it has been that* (*both*

[1] For solutions available on other assumptions, see my *Past, Present and Future* (Oxford 1967), pp. 190–5, and *Papers on Time and Tense* (Oxford 1968), p. 129.

[2] In *Past, Present and Future*, pp. 89–90, and *Papers on Time and Tense*, pp. 101, 126, 138, I use the equivalent form 'At all times, *if A then Brown is ill*'.

A and Brown is ill) or it will be that (both A and Brown is ill).

And finally,

(40) The instant *a* is earlier than the instant *b*
= (41) That *it will be that B* is the case at *a*
= (42) At some time, *both A and it will be that B*
= (43) *Either (both A and it will be that B) or it has been that (both A and it will be that B) or it will be that (both A and it will be that B).*

In passing from (31) to (32), from (33) to (36), from (37) to (39) and from (40) to (43) we start with propositions of our untensed 'exterior' language and proceed stage by stage to equivalent propositions of our tensed 'interior' language, until the latter encompasses the whole.

4. *Derivation of other modes of individual reference within Egocentric*
We may similarly build up non-egocentric modes of individual reference within Egocentric on the Leibnizian assumptions that individuals form a linear series in the order of their perfection, and that each individual has something which is true of him only, i.e. an egocentric proposition which is true only when *he* says it. Philosophically the most interesting proposition which is true of a given individual only is the conjunction of all the truths that concern him, but for formal purposes any proposition which is true of him only, i.e any egocentric proposition which is true only when he says it, will do. As Castañeda has nicely put it, each of us can say in Egocentric, 'I am a true proposition and everyone else is a false one.' For example, if I am represented by the proposition *Priorising*, Brown by *Brownising*, etc., I can truly say 'Priorising but not Brownising and not Quinising, etc.', i.e. 'I am Prior but I am not Brown and not Quine, etc.'. We now have these equivalences:

(44) I alone Priorise
= (45) *Priorising* is the case with me only
= (46) *Priorising but not Inferior-to-Priorising and not Superior-to-Priorising.*

And

(47) Only one individual Priorises
= (48) *Priorising* is the case with only one individual
= (49) It is the case with someone that (46)
= (50) Either it is the case with me, or it is the case with someone more perfect, or it is the case with someone less less perfect, that (46)
= (51) *Either* (46) *or Inferior-to-*(46) *or Superior-to-*(46).

Again

(52) Brown is standing
= (53) *Standing* is the case with Brown
= (54) *Brownising and standing* is the case with someone
= (55) *Brownising and standing* is the case either with me or with someone more perfect or with someone less perfect
= (56) *Either* (*Brownising and standing*) *or Inferior-to-*(*Brownising and standing*) *or Superior-to-*(*Brownising and standing*).[1]

And finally

(57) Brown is less perfect than Quine
= (58) *Inferior-to-Quinising* is the case with Brown
= (59) *Brownising and inferior to Quinising* is the case with someone
= (60) *Either* (*Brownising and inferior to Quinising*) *or Inferior to* (*Brownising and inferior to Quinising*) *or Superior to* (*Brownising and inferior to Quinising*).

In passing from (44) to (46), from (47) to (51), from (52) to (56), and from (57) to (60) we begin with our ordinary ways of describing individuals, and finish with something in Egocentric. Each of our end-points (46), (51), (56), (60) is in fact an implicit characterisation of *the speaker*. For example,

(56) = (61) *I* am either Brownising and standing or am less perfect than someone who is doing so or am more perfect than someone who is doing so.

[1] For a modification necessary with less simple assumptions see *Papers on Time and Tense*, p. 140.

In this way a typical statement about Brown is reduced to one about me, and in a language with no other subject than me this one subject of all that I say can go unmentioned.

5. *Egocentric and the logic of personal pronouns*

Egocentric is not a very promising stylised language for the development of a logic of personal pronouns, since it is distinguished precisely by leaving these out. Nevertheless, some of the problems which do arise in understanding personal pronouns may be illuminated by translating certain English sentences into Egocentric. H.-N. Castañeda[1] has recently drawn attention to certain peculiarities of such inferences as

(62) Brown knows that he is ill; therefore he is ill.

This seems at first a simple instance of the valid inferential form '*x* knows that *p*, therefore *p*'. However, as Castañeda points out, the form of words 'He is ill' is here being used very differently in the premiss and the conclusion. In the premiss, 'Brown knows that he is ill', the fragment 'he is ill' expresses Brown's *self*-knowledge, i.e. it expresses what Brown would himself express by saying '*I* am ill'. The premiss does not mean the same as 'Brown knows that Brown is ill', for Brown might know that *he* was ill even if through some unhappy accident he did not know that he was Brown. In the conclusion, however, we say 'He is ill', not to express any sort of self-knowledge, but just as a short way of saying that *Brown* is ill. If we used the word 'Self' to mean 'I' in a sentence on its own and the first sort of 'He' in reporting the self-knowledge of someone other than the speaker, it would seem that the inference

(63) Brown knows that self is ill, therefore self is ill

would genuinely be of the form '*x* knows that *p*, therefore *p*', but would not be valid, since it would mean 'Brown knows that he is ill, therefore *I* am ill'; whereas the inference

(64) Brown knows that self is ill; therefore Brown is ill

[1] For example, in 'The Logic of Self-Knowledge', *Nous*, 1 (1967).

would give the force of the original inference, and would be valid, but would not be of the form '*x* knows that *p*, therefore *p*'. On the other hand,

(65) Self knows that self is ill, therefore self is ill,

i.e. 'I know that I am ill, therefore I am ill' would be both valid and of the form in question. If, now, we translate (63), (64) and (65) into Egocentric (using *Someone p for p or inferior-to-p or superior-to-p*), we get

(66) *Someone Brownising and knowing being ill, therefore being ill*
(67) *Someone Brownising and knowing being ill, therefore someone Brownising and being ill*
(68) *Knowing being ill, therefore being ill.*

Here the valid (68) alone is of the form 'Knowing *p*, therefore *p*'; (67) is still not of this form, but its validity is easily derivable from the validity of this form, and the invalid (66) is not of this form at all; in fact, *all* inferences that are *really* of the form 'Knowing *p*, therefore *p*', or of derivable forms, *are* valid when we have got them into this language.

6. *Egocentric as a key to the philosophy of Leibniz*

On the subject of *temporal* reference, the development of the logic of the earlier-later relation as a simple extension of tense logic, i.e. the development of which a simplified form is sketched in Section 3, seems to me much more than a symbolic dodge, for I find myself quite unable to take 'instants' seriously as individual entities; I cannot *understand* 'instants', and the earlier-later relation that is supposed to hold between them, except as logical constructions out of tensed facts. Tense logic is for me, if I may use the phrase, *metaphysically fundamental*, and not just an artificially torn-off fragment of the first-order theory of the earlier-later relation. Egocentric logic is a different matter; I find it hard to believe that individuals really are just propositions of a certain sort, or just 'points of view', or that the real world of individuals is just a logical construction out of such points of view. Nevertheless the fact that we can have a consistent and comprehensive egocentric logic as well as a logic of tenses does suggest that some sort of idealism or relativism is a more defensible philosophical position than it once looked. And in particular, when I drew upon certain elements from the philosophy of Leibniz when working out the details of Egocentric, this was not an arbitrary or accidental choice; it does seem to me that much in the work of that philosopher takes on a new significance when we think of him as a man who would have regarded the 'egocentric' account of the world as 'metaphysically fundamental'. I conclude by mentioning some detailed points which support this suggestion.

(a) For Leibniz, self-knowledge was the starting-point of his understanding of the world. He said, for example, 'Since I conceive that other beings have also the right to say *I*, or that it may be said for them, it is by this means that I conceive what is called substance in general.'[1] The parenthetical 'or that it may be said for them' is worth noting. In relating Egocentric to more 'objective' languages, I have constantly referred to *the speaker*; but no such person is mentioned within Egocentric itself, and the facts which may be stated in Egocentric (e.g. what I remember when I remember *being at the meeting*) might also go *un*stated; e.g. the fact which my pencil would express, if it could think or talk, by saying 'Someone more perfect than me is holding me'.

(b) 'In consulting the notion which I have of every true proposition,' Leibniz says in a notorious passage, 'I find that every predicate, necessary or contingent, past, present or future, is comprised in the notion of the subject.'[2] So a subject is a conjunction of predicates, i.e. a compound egocentric *proposition*.

(c) But the most fundamental point is this: In tense-logic the totalities of tensed propositions which are true at different instants fit together into a *system*, so that although the total course of history will be differently described at different times, the description at one time will determine what the descriptions at other times will be. For example, because a certain past totality of truth included the proposition that the Battle of Hastings *is* occurring, the present totality of truth includes the proposition that the Battle of Hastings *was* occurring. Similarly the world as a whole will be differently described by different people using an egocentric language, but how

[1] Cited in Russell's *Philosophy of Leibniz*, p. 215.

[2] Ibid., p. 206.

it is described by one person will determine how it is described by another. For example, if *Inferior-to-drinking* or *Someone-more-perfect drinking* is how things are with me, then just *Drinking* ('I am drinking') is how things are with someone more perfect than me; and because *Drinking* is part of the totality of truth that makes up his being, *Someone-more-perfect drinking* is part of the totality that makes up *my* being. This, of course, is the 'pre-established harmony'; whereby although 'every soul is as a world apart', yet 'each substance expresses the whole sequence of the universe according to the view or respect which is proper to it'.[1]

(d) Leibniz notoriously had no place for genuine *relations* between individuals. 'Paternity in David is one thing, and filiation in Solomon is another, but the relation common to both is a merely mental thing.'[2] In Egocentric, two-place predicates like 'is less perfect than' disappear into modalisings of propositions, with one modalising of a proposition in one personal totality requiring an appropriate other modalising in another personal totality.

(e) Leibniz wavered in his egocentricity when talking about God. 'Every soul is a world apart' continues: 'independent of everything else but God,' suggesting that *God*'s egocentricity is somehow not egocentric at all. This won't do. I once said that if God did not see the past as past, he would be unaware of the fact at which I am rejoicing when I say 'Thank goodness that's over', but only of the quite un-gratifying fact that something is earlier than something else; Mr. Anselm Müller suggested that if these were distinct 'facts' so would be the fact at which I rejoice when I am glad at *having won a bet* and the person-neutral *Someone Priorising and having won a bet* (a conjunction at which I have no particular reason to be pleased); but surely, Müller argued, it is no limitation to God's omniscience that the only fact (of these two) that he can see is the latter. (For God, *having won a bet* is not a fact at all, since not he but I won it.) Well, a person, e.g. Leibniz, who took Egocentric as seriously as I take tenses would have to say that *having won a bet* really *is* a fact which God cannot see as one, just as no one can see a war as being over until it *is* over.[3] But Leibniz seems to try,

[1] Ibid., p. 206.

[2] Ibid., p. 207.

[3] Cf. H.-N. Castañeda, 'Omniscience and Indexical Reference', *Journal of Philosophy*, LXIV (April 1967).

inconsistently, to put God outside this relativity. McTaggart is a more consistent Leibnizian at this point.

(f) However, there is no reason why there should not be a 'Supreme Monad', i.e. one than which none is more perfect. (McTaggart too seems to admit this.[1]) There is a tense-logic for a time which ends; one of its laws is 'For any *p*, either it will not be the case that *p* (because it is already the end of time and nothing "will be the case") or it will be that it will not be the case that *p* (because it will be that nothing further "will be the case")'. An egocentric logic with a Supreme Monad would analogously contain the law 'For any *p*, either not-inferior-to-*p* (because I am the Supreme Monad) or inferior-to-not-inferior-to-*p* (because someone more perfect than me is the Supreme Monad)'. This is, in fact, precisely how the existence of a Supreme Monad would have to be stated in Egocentric.

Egocentric is also relevant to some more modern philosophising. I am never quite sure what is being affirmed or denied when people debate the possibility of a private language; but there is surely *some* sense of 'private language' in which Egocentric is one, although it is a language in which we can quite effectively communicate with one another because of the correspondences mentioned above under (c). Moreover, it offers a consistent development of the view that no one is directly acquainted with, or can directly talk about, any 'particular' but himself; our derivation, in Section 4, of ordinary ways of speaking within egocentric ones suggests how something like a common world *can* be built up from essentially private experience. Egocentric logic could also be exploited in defence of the Wittgensteinian thesis (if it is one) that 'I am in pain' has a *different logical form* from 'Brown is in pain'. If *this* is the thesis, then it is more responsible than I thought at first.

7. *Modality, quantification and individuality in Peirce*

There are also hints of some of the above lines of thought in C. S. Peirce.

(a) Peirce has a persistent habit of treating quantification as a special sort of modality, and quantification over individuals as a special sort of quantification over states of affairs, rather than *vice versa*. We encounter this subsumption of individuals under 'worlds',

[1] See, e.g., *The Nature of Existence*, vol. II, pp. 182–3.

for example, in one of his expositions of his 'logical graphs', in which conjunction is expressed by juxtaposition and negation by bracketing, i.e. he uses *AB* for '*A* and *B*' and '(*A*)' for 'Not *A*'. A nice notation. We can read such diagrams, he says, in two ways. In the simplest way we take them to relate to 'a single state of the universe, like the present instant'. This is in effect how current tense-logic proceeds. But we may also read an unmodified proposition as meaning 'that it is sometimes or possibly true', so that the compound *AB* signifies that *A* and *B* are each of them 'at some quasi-instant true', and (*A*) and *A* is true at no quasi-instant. Peirce then enriches the symbolism, taken this second way, not with names of particular states of affairs or instants, but with 'lines of identity', so that *A–B* means that *A* is true at some instant and *B* at *the same* instant ('Simultaneously *A* and *B*'), or that they are true in some *the same* possible state of affairs, i.e. that they are 'compossible'. Peirce then says—and this is, for our present purpose, the 'punch line'—that 'our quasi-instants may be individuals', in which special case his second reading equates *A* with 'Something is *A*', (*A*) with 'Nothing is *A*', *AB* with 'Something is *A* and something is *B*', *A–B* with 'Something is both *A* and *B*', and so on.[1]

(b) Elsewhere Peirce suggests that individual terms are just general terms with a particular special feature. 'Individuals are either identical or mutually exclusive, and cannot intersect or be subordinated to one another as classes can'. 'The logical atom, or term not capable of logical division, must be one of which every predicate may be universally affirmed or denied. For example, let *A* be such a term. Then if it is neither true that all *A* is *X* nor that no *A* is *X*, it must be that some *A* is *X* and some *A* is not *X*, and therefore *A* may be divided into *A* that is *X* and *A* that is not *X*, which is contrary to its nature as a logical atom.'[2] This reduces the relation of a predicate to its subject, to a relation of a predicate to another predicate of a particular sort; and predicates are notoriously, for Peirce, *just slightly damaged propositions*,[3] so we have our own theory. And as to the particular feature of 'atomic' terms which

[1] C. S. Peirce, Collected Papers, 4.376, 385. For a formal system based on these diagrams, see J. Jay Zeman, 'A System of Implicit Quantification', *Journal of Symbolic Logic*, XXXII (December 1967), pp. 480–504.

[2] Ibid., 3.92–3.

[3] Ibid., 2.344, 4.572.

Peirce singles out, it is like saying in tense-logic that if the tensed proposition *A* is an *instant*, then for any *X*, either *A* at all times implies *X* or *A* at all times implies not *X*; this is in fact provable in linear tense-logic when 'instant' and 'always' (i.e. 'not sometimes not') are defined as in our (32) and (36) above.[1] Similarly it should result from adequate egocentric definitions of 'individual' and of 'everyone' that if *A*-ing is an individual or individuating proposition, then for any *X*, either it is true of everyone that (his) *A*-ing implies (his) *X*-ing, or it is true of everyone that (his) *A*-ing implies (his) not *X*-ing; e.g. either everyone who Brownises is ill or everyone who Brownises is not ill.[2]

(c) After giving this account of a 'logical atom', Peirce goes on immediately to say that 'such a term can be realised neither in thought nor in sense', and that 'absolute individuality is merely ideal'. Not that we cannot guarantee terms that satisfy the condition of atomicity just mentioned, for any *empty* term will satisfy it trivially (if there are no *A*'s, then *both* all *A*'s are *X* and no *A*'s are *X*, whatever *X* may be); but non-emptiness is another condition of individuality, and we cannot guarantee that there are any terms that satisfy this and Peirce's condition as well. At all events, it is possible to produce a modal theory of *possible-world-propositions*, a tense-logical theory of instant-propositions, or an egocentric theory of individual-propositions, which does not assume that any actual proposition satisfies the definition of an individual or instant or state-of-the-world. I conclude by sketching a simple calculus of this sort, in three versions.

8. *A calculus of worlds* (*instants, individuals*)

We use the Łukasiewicz symbols *C*, *K*, *A*, *N*, *E* for 'if', 'and', 'or', 'not', 'iff', and use *Lp* for 'Everywhere *p*' and *Mp* (= *NLNp*) for 'Somewhere *p*', with the S5 laws:

RL: If $\vdash\alpha$ then $\vdash L\alpha$
L1. *CLCpqCLpLq*
L2. *CLpp*
L3. *CNLpLNLp*.

[1] Cf. *Past, Present and Future*, p. 83.
[2] Cf. *Papers on Time and Tense*, pp. 140–1.

For 'p is an individual' (or an instant, or a possible total world-state) we write Qp. If we have propositional quantifiers, we can define Qp thus:

$$Qp = KMp\Pi qALCpqLCpNq$$

(Peirce's definition above, with the added conjunct Mp to exclude mere impossibilities.) Using this, we can derive in quantified S5 the following:

Q 1. $CQpMp$
Q 2. $CQpALCpqLCpNq$
Q 3. $CQpLQp$
Q 4. $CLCpqCLCqpCQpQq$.

Alternatively we may do without quantifiers and take Q as undefined with Q 1–4 as its characteristic axioms. They are independent (to prove this for Q 1 let $Q = NM$; for Q 2, let $Q = M$; for Q 3, let $Q\alpha = KQ\alpha\alpha$; for Q 4—I owe this to Mr. K. Fine—use I for perfect identity, k for some constant, and read $Q\alpha$ as $KQ\alpha I\alpha k$). We can, however, replace Q 1 + Q 4 by the one axiom $CLCpqCLCqpCQpLQq$ (also due to K. Fine). If we just add propositional quantifiers with the usual rules, we cannot prove from this basis

Q 5. $CMpC\Pi qALCpqLCpNqQp$

(proof of independence: let $L\alpha = \alpha$ and $Q\alpha = NC\alpha\alpha$; K. Fine), which with Q 1 and Q 2 would be equivalent to the definition of Q in Π and L. But I do not know of any *quantifier-free* consequence of Q 1–5 which does not also follow from Q 1–4. For example, from Q 1–4 we can prove $CQpELCpqNLCpNq$, $CMQpQp$, $CQQpEQqLq$ and $CQQpEqLq$ (i.e. if Qp is itself a world, instant or individual, there is only one such).

If we use Op for $\Pi qALCpqLCpNq$, i.e. for the 'L-completeness' which worlds (instants, individuals) share with impossibilities, we can define O in a Q-primitive system as $ANMpQp$, while in an O-primitive system we could define Qp as $KMpOp$ and axiomatise by

O1. $CLNpOp$
O2. $COpALCpqLCpNq$
O3. $COpLOp$
O4. $CLCpqCOqOp$

(where O1 is independent by $O\alpha = NC\alpha\alpha$ (J. T. Canty), O2 by $O\alpha = C\alpha\alpha$ (Canty), O3 by $O\alpha = ALN\alpha K\alpha O\alpha$; O4 as Q 4; but O3 + O4 may be replaced by $CLCpqCOqLOp$). These are equivalent to Q 1–4, and are to

O5. $C\Pi qALCpqLCpNqOp$

as those are to Q 5.

We can also axiomatise in W, where $Wp = KpQp$ ($KpOp$), meaning that p is the *actual* world, the *present* instant, the individual that I am. Q would be definable in this system as MW, and O in terms of Q as before. For axioms (yielding an equivalent system) we need only

W1. $CWpp$
W2. $CWpCqLCpq$
W3. $CLCpqCLCqpCWpWq$

(where W1 is independent by $W = LN$; W2 by $W\alpha = \alpha$ (Canty); W3 as Q 4). These are to

W4. $CpC\Pi qCqLCpqWq$

as Q (O) 1–4 are to Q 5 (O5).

One formula which does not follow from W1–3 (Q 1–5; O1–5) is

W5. ΣpWp (= Q 6. $\Sigma pKpQp$ = O6. $\Sigma pKpOp$)

asserting that there is an actual world (present instant, individual that I am). Indeed, even ΣpQp is not derivable from this basis, though $OKpNp$, and so ΣpOp, follows easily enough from O1 and $LNKpNp$. There is, certainly, no *finite* model satisfying W1–4 (Q 1–5; O1–5) which would refute W5 (Q 6; O6), since we can prove for any string of variables p, q, r, . . ., the thesis

$$CpCqCr \ldots \Sigma sKsKLCspKLCsqKLCsr \ldots,$$

i.e. given any finite list of truths we can find a truth (namely their conjunction) which everywhere (always, necessarily) implies them all; and if these and their consequences were all the truths there are their conjunction would satisfy the definition of *W* and so W1–4. But with an infinity of truths there might be no such conjunction. (To refute the weaker ΣpQp we need only suppose that things are *always* like this.) There is perhaps a criticism of Descartes to be drawn from this. From 'I am thinking', i.e. 'Thinking' (p), egocentric logic does enable us to infer 'Thinking somewhere' (Mp), but not 'Thinking in some *individual*' ($\Sigma qKQqLCqp$), at least not unless 'There is an individual that I am' (Q6) is laid down as a special axiom.

When W5 (Q6; O6) is added to W1–3 (Q1–4; O1–4), it becomes possible to prove W4 (Q5; O5). In *W* the proof is

$C(1)p$	
$C(2)\Pi qCqLCpq$	
$\Sigma rK(3)Wr$	(W5)
$K(5)r$	(3, W1)
$K(6)LCpr$	(2, 5)
$K(7)LCrp$	(3, W2, 1)
$(8)Wp$	(6, 7, W3).

The addition of the weaker ΣpQp to Q1–4 (O1–4; W1–3) does not yield Q5 (O5; W4). (For independence, let n be the actual world and read $Q\alpha$ as $LE\alpha n$; K. Fine.) But I do not know of any *quantifier-free* consequence of Q1–4 (O1–4; W1–3) with even the stronger W5 (Q6; O6) added which does not follow from the Q1–4 (O1–4; W1–3) alone; it would seem that the actual existence of worlds (instants, individuals) cannot be reflected in a postulate of the unquantified system.

3. *Supplement to 'Egocentric Logic'*

1.1. Consider the variable-binding operator for 'exactly one thing'. We would write it as *Qx*, so that *Qxpx* would mean 'Exactly one thing is *p*', and '*The* thing which is *p* is *q*' could be written as *KQxpxMxKpxqx*.

If we enrich the uniform monadic first-order predicate calculus with this special quantifier, it will have to have its own special postulates; and it is now known which these must be, namely anything equivalent to the following form axiom-schemata:

Q 1. $CQx\alpha Mx\alpha$
Q 2. $CQx\alpha ALxC\alpha\beta LxC\alpha N\beta$
Q 3. $CQx\alpha LxQx\alpha$
Q 4. $CLxE\alpha\beta CQx\alpha Qx\beta$.

The sufficiency of these (with the *x*'s deleted, but it is at this point a matter of indifference whether we operate as in 1.3 or as in 1.4) was conjectured by me in 'Egocentric Logic' (Chapter 2 above) and was shortly afterwards established independently by K. Fine, R. A. Bull and D. Kaplan. Q 1 states that what is true of exactly one individual is true of at least one individual; Q 2 that if exactly one thing satisfied α, then either everything that satisfies α satisfies β or nothing that satisfied α satisfies β; Q 3 that if α is true of exactly one individual, then it is true of *every* individual *that* α is true of exactly one; Q 4 that if α and β are true of exactly the same individuals, then if α is true of exactly one individual, so is β.

1.2. It is a commonplace that of the quantifiers *Lx* and *Mx*, either may be taken as primitive and the other defined in terms of it (*Mx* as *NLxN*, or *Lx* as *NMxN*), the postulates being adjusted to the quantifiers taken as primitive. Our special individualising quantifier Q may similarly be either defined in terms of something else, or used to define that something else. For example, we could introduce the form $Wx\alpha$ for 'I and I alone satisfy α', and define $Qx\alpha$ as $MxWx\alpha$, i.e. as 'There is some individual who can say truly, "I and I alone satisfy α" '. Alternatively we could take Q, as before, as undefined, and define $Wx\alpha$ as $K\alpha Qx\alpha$, i.e. 'I and I alone satisfy α' as 'I satisfy α, and α is satisfied by exactly one individual'. In 'Egocentric Logic' I have given the following postulates for W:

W1. $CW\alpha\alpha$
W2. $CW\alpha C\beta LC\alpha\beta$
W3. $CLE\alpha\beta CW\alpha W\beta$.

W1 asserts that what is true of me only is true of me, or that which is true now only is true now, or that what is true in the natural world only is true. W2 asserts that if I alone satisfy α, then if I satisfy β, then everyone satisfies 'If α then β' (I do because *ex hypothesi* I satisfy β, and everyone else because *ex hypothesi* other people than me don't satisfy α, and so do satisfy 'If α then β'). Or in tense logic it asserts that if it is now but at no other time the case that α, and now the case that β, then it is always the case that if α then β; or in modal logic, that what is the case in the actual world only necessarily implies whatever is the case in this world. W3 asserts that if α and β are satisfied by exactly the same individuals, then if I and I alone satisfy α, I and I alone satisfy β (the other interpretations are obvious).

1.3. I asserted but did not prove above that W1–3, with Q defined as MW, was equivalent to Q1–4, with $W\alpha$ defined as $K\alpha Q\alpha$. I now prove the former and more difficult of these (citing theorems of S5 without proof):

1. $CMW\alpha M\alpha$ (W1; $\vdash C\alpha\beta \rightarrow CM\alpha M\beta$)
Q1. $CQ\alpha M\alpha$ (1, Df. Q)
2. $CW\alpha CN\beta LC\alpha N\beta$ (W2)
3. $CW\alpha ALC\alpha\beta LC\alpha N\beta$ (W2, 2, $A\beta N\beta$)
4. $CMW\alpha MALC\alpha\beta LC\alpha N\beta$ (3, $\vdash C\alpha\beta \rightarrow \vdash CM\alpha M\beta$)
5. $CMW\alpha AMLC\alpha\beta MLC\alpha N\beta$ (4, $MA = AM$)
6. $CMW\alpha ALC\beta LC\alpha N\beta$ (5, $ML = L$)

Q 2. $CQ\alpha ALC\alpha\beta LC\alpha N\beta$ (6, Df. Q)
7. $CMW\alpha LMW\alpha$ ($M - LM$)
Q 3. $CQ\alpha LQ\alpha$ (7, Df. Q)
8. $CLE\alpha\beta LCW\alpha W\beta$ (W3, RL)
9. $CLE\alpha\beta CMW\alpha MW\beta$ (8, $CLC\alpha\beta CM\alpha M\beta$)
Q 4. $CLE\alpha\beta CQ\alpha Q\beta$ (9, Df. Q)
10. $CC\alpha NW\alpha CW\alpha NW\alpha$ (W1, syll.)
11. $CC\alpha NW\alpha NW\alpha$ (10, $CC\alpha N\alpha N\alpha$)
12. $CLC\alpha NW\alpha LNW\alpha$ (11, $\vdash C\alpha\beta \rightarrow \vdash CL\alpha L\beta$)
13. $CNLNW\alpha NLC\alpha NW\alpha$ (12, transp.)
14. $CQ\alpha NLC\alpha NW\alpha$ (13, Df. M, Df. Q)
15. $CQ\alpha LC\alpha W\alpha$ (14, Q 2)
16. $CK\alpha Q\alpha W\alpha$ (15)
17. $CW\alpha K\alpha MW\alpha$ (W1; $C\alpha M\alpha$)
18. $CW\alpha K\alpha Q\alpha$ (17, Df. Q)
19. $EW\alpha K\alpha Q\alpha$ (16, 18).

This last equivalence corresponds to the definition of W in the Q-primitive system.

2.1. Considered as variable-binding operators, Q and W differ in important respects both from L and M and from one another. Let us take Q first. By axiom Q 3, $Q = LQ$ ($Qx = LxQx$). When Lx is prefixed to Qx, as when it is prefixed to Mx or Lx, it becomes a 'vacuous' quantifier, binding a variable that is already bound. One could, in fact, just drop Q 3, $CQx\alpha LxQx\alpha$, and regain it from $CQx\alpha Qx\alpha$ and RL, if it is understood that a variable within the scope of a Q, like one within the scope of an L, is not free. On the other hand, prefixing Qx to a formula that is already 'closed' is not by any means a vacuous operation. For if α is already closed (i.e. its x is nowhere free) then $Qx\alpha$ is in general a much stronger assertion than α, and will entail, whether α already entails this or not, that there is only one individual in the universe. We can establish this conclusion in various ways; for example thus: if α is already closed, and $Qx\alpha$ is true, then either α is itself true, or it is not. If it is true, it is (vacuously) satisfied by all the individuals there are, i.e. $Lx\alpha$ is also true. But if α is satisfied by all the individuals there are, as $Lx\alpha$ asserts, and also by exactly one individual, as $Qx\alpha$ asserts, it immediately follows that exactly one individual is all the individuals there are. If on the other hand, the closed formula α is false, then it is not satisfied by any individual, i.e. we now have 'For no *x*, α'; but in this case $Qx\alpha$, asserting 'For exactly one *x*, α', will also be false. So where α is closed, $Qx\alpha$ is only true when there is just one individual. If we go outside our present calculus and define $Qx\alpha$ as $MxLyE\alpha Iyx$, we can easily formalise the above reasoning. If α is true, $Qx\alpha$ will assert that for a certain *x*, every *y*'s identity with this *x* is equivalent to that true proposition, i.e. every *y* is identified with this *x*. If α is false, $Qx\alpha$ will assert that for a certain *x*, every *y*'s identity with this *x* is equivalent to that false proposition; i.e. there is an *x* of which it is false to say that anything is identical with it. But there is no such *x* (since every *x* is identical with itself), so in this case, i.e. in the case where α is false, $Qx\alpha$ will be false too. And within our calculus, we can prove that if α is closed and so is equivalent to $Lx\alpha$, then $Qx\alpha$ implies $C\beta Lx\beta$, i.e. whatever is true of me is true of everything, i.e. there is no one but me. Once again in giving the proof it is simplest to drop the *x*'s; we then have

$C(1)LC\alpha L\alpha$ (closure of α)
$C(2)Q\alpha$
$C(3)\beta$
$K(4)ALC\alpha\beta LC\alpha N\beta$ (2, Q 2)
$K(5)ACL\alpha L\beta CL\alpha LN\beta$ (4, $CLCpqCLpLq$)
$K(6)M\alpha$ (2, Q 1)
$K(7)ML\alpha$ (1, 6, $CLCpqCMpMq$)
$K(8)$ (8, 7, $ML = L$)
$K(9)AL\beta LN\beta$ (5, 8)
$K(10)NLN\beta$ (3)
$(11)L\beta$ (9, 10).

Hence, since $Qxpx$ and $Lxpx$ are closed (or: QQp and Lp fully modalised), we have as special cases of what has just been proved $CQxQxpxCqLxqx$ ($CQQpCqLq$) and $CQxLxpxCqLxqx$ ($CQLpCQqLq$).

2.2. With *W*, we have a very different story. Whereas with *Q* we have $EQ\alpha LQ\alpha$ but not $EQ\alpha QQ\alpha$, with *W* we don't have $EW\alpha LW\alpha$ but do have $EW\alpha WW\alpha$. If we recall that $W\alpha$ is equivalent to

$K\alpha Q\alpha$, and $Wx\alpha$ to $K\alpha Qx\alpha$, these results will not be wholly surprising. The essential point is that $Wxpx$, being equivalent to $KpxQxpx$, is not strictly speaking a closed formula, but merely contains one. In 'I and I alone can p', re-stated as 'I am p and exactly one individual is p', only the second conjunct is quantified. Similarly in modal logic, where Wp may be read as 'p is true in the actual world only', or in tense logic, where Wp may be read as 'p is true now but at no other time', when we equate this with $KpQp$ we see that it is not a 'fully modalised' formula, even if Qp is. $LxWxpx$ would mean 'Everyone can truly say "I and I only am p" ', and this does not follow from *my* being able to say this truly; at least, it would only follow if I were the only individual. To lay down $CWx\alpha LxWx\alpha$ as a law would thus be a way of saying that there is no individual but me. On the other hand, if I and I alone am p, it does follow that I and I alone can truly say 'I and I alone am p'; that is, from 'I am p and exactly one individual is p' I can infer the conjunction of this with 'Exactly one individual can truly say: I am p and exactly one individual is p'. Or in symbols (dropping x's) $CKpQpKKpQpQKpQp$, or $CWpWKpQp$, i.e. $CWpWWp$. Directly using our postulates for W, we can prove this quite generally as follows:

1. $CW\alpha CW\alpha LC\alpha W\alpha$ (W2)
2. $CW\alpha LC\alpha W\alpha$ (1, $CC\alpha C\alpha\beta C\alpha\beta$)
3. $LCW\alpha\alpha$ (W1, S5)
4. $CW\alpha LE\alpha W\alpha$ (2, 3)
5. $CW\alpha CW\alpha WW\alpha$ (4, W3).

4. *Worlds, Times and Selves*

1.1. In what follows I shall use the Łukasiewicz symbolism for propositional calculus, i.e. *Np* for 'Not *p*', *Cpq* for 'If *p* then *q*', *Apq* for 'Either *p* or *q*', *Kpq* for 'Both *p* and *q*' and *Epq* for 'If and only if *p* then *q*'. To these are added *Lp* for the modal function 'Necessarily *p*', and *Mp* for 'Possibly *p*'. In this symbolism we may axiomatise the modal system S5 by subjoining to all tautologies the special axiom-schema

A1. $CL\alpha\alpha$

and to the rule of detachment (*modus ponens*) the special rule

RL: $\vdash C\alpha\beta \rightarrow \vdash C\alpha L\beta$,

for every α that is 'fully modalised' (i.e. each of its propositional variables falls within the scope of some occurrence of *L*). *M* ('possibly') is defined as *NLN* ('Not necessarily not'). That these postulates yield exactly S5 was shown by Lemmon.[1]

1.2. It is well known that S5 may be interpreted as a fragment of ordinary predicate calculus. To make this interpretation obvious, we may construct the relevant fragment of predicate calculus as follows: We employ a single individual variable *x*, and an unlimited set of predicate variables *p*, *q*, *r*, etc. If φ is any predicate variable, φx is a well-formed formula. If α and β are well-formed formulae, so are $C\alpha\beta$, $A\alpha\beta$, $K\alpha\beta$, $E\alpha\beta$, $N\alpha$, $Lx\alpha$ and $Mx\alpha$. $Lx\alpha$ is read as 'For all *x*, α' and $Mx\alpha$ as 'For some *x*, α'. Typical well-formed formulae are

[1] E. J. Lemmon, 'Alternative Postulate-sets for Lewis's S5', *Journal of Symbolic Logic*, vol. 21, no. 4 (December 1956).

px, meaning '*x* is *p*'.
Npx, meaning '*x* is not *p*'
Cpxqx, meaning 'If *x* is *p* then *x* is *q*'
Lxpx, meaning 'Everything is *p*'
LxCpxqx, meaning 'Everything that is *p* is *q*'.

We may axiomatise the system by subjoining to all tautologies the axiom-schema

A1′. $CLx\alpha\alpha$

and to the rule of detachment the rule

RL′: $\vdash C\alpha\beta \rightarrow \vdash C\alpha Lx\beta$,

for every α in which *x* is not free. *Mx* is defined as *NLxN*. We may call this the uniform monadic first-order predicate calculus. It is a system in which we can make statements that are either about everything (where *x* is bound), or about a particular privileged individual *x* (where *x* is free). There may be other individuals, but only this one can be referred to individually. It is clear that in this calculus the symbol *x* is not really needed, but can be omitted as 'understood' throughout. If we thus erase it, we may then read

p as '*x* is *p*'
Np as '*x* is not *p*'
Cpq as 'If *x* is *p* then *x* is *q*'
Lp as 'Everything is *p*'

and so on. And when thus re-written, the system simply becomes S5 as previously formulated.

1.3. Just what is going on here? What, in other words, do I mean by calling this an 'interpretation' of S5? In the first instance, just this: I have the symbolic calculus described in 1.1. Where did it come from? In fact, of course, it came from an attempt to set out symbolically the logic of necessity and possibility. But I could have obtained exactly the same symbolic calculus in quite a different way, namely by taking a certain fragment of predicate calculus, i.e. of the logic of 'all' and 'some', and symbolising it a little curiously—first writing *p*, *q*, *r*, etc. instead of *f*, *g*, *h*, etc. for predicates (what's so sacred about the shapes *f*, *g*, *h*, after all?), and *Lx* instead of Π*x* for the universal quantifier (what's so sacred about the shape Π, after all?); and then deleting all occurrences of the name-variable *x*, which in this fragment of the predicate calculus I can safely do because no other name-variables occur. And then, behold, I have precisely the calculus described in 1.1.

2.1. These are, to be sure, *different* ways of arriving at the calculus described in 1.1. To introduce *p*, *q*, *r*, etc. as variables standing for propositions is one thing, to introduce them as variables standing for predicates is quite another thing; and again, to introduce *L* as a constant meaning 'necessarily' is one thing, and to introduce it as a constant meaning 'For all ——' is quite another thing. *Quite* another thing? Suppose I arrive at the calculus by the procedure of 1.2, only I don't let my *x*'s stand for any old individuals but for individuals of a very special sort, namely possible states of affairs or 'worlds'; the privileged individual represented by the one variable *x* when free being the *actual* world. Similarly, I don't let my *p*'s, *q*'s etc. stand for any old predicates but for predicates of the type '— has the proposition that *p* true in it'; i.e. *px* means 'The world *x* has the proposition that *p* true in it', i.e. 'It is the case in the world *x* that *p*' (a predicate eminently suitable for this sort of subject). Then *px*, with *x* free, will mean 'It is the case in the actual world that *p*', i.e. 'It actually is the case that *p*', i.e. 'It *is* the case that *p*'; i.e. *p*. And *Lxpx* will mean 'Every possible world has *p* true in it', i.e. 'It is the case in every possible world that *p*'; and this, according to Leibniz, is precisely what we mean when we say 'Necessarily *p*'.

2.2. So the original, normal or standard interpretation of the calculus sketched in 1.1, i.e. the interpretation of it as a logic of necessity and possibility, can be presented as just a special case of the interpretation of it as a mildly odd formulation of the uniform monadic lower predicate calculus. It *can* be so presented. But do we illuminate the subject of modal logic by so presenting it? To this I want to say, No; or at all events, Not much. It is, if you like, formally but not materially illuminating to present modal logic thus. The metatheory of predicate calculus is more fully understood than that of modal logic, so that the presentation of the latter as a special case of the former enables certain transfers of information to take place. But possible worlds, in the sense of possible states of affairs, are not *really* individuals (just as numbers are not *really* individuals). To say that a state of affairs obtains is just to say that something is the case; to say that something is a possible state of affairs is just to say that something could be the case; and to say that something is the case 'in' a possible state of affairs is just to say that the thing in question would necessarily be the case if that state of affairs obtained, i.e. if something else were the case. That is, the proper logical form of the statement that it is true in the state of affairs *p* and *q*, is just: 'Necessarily if *p* then *q*', *LCpq*. We understand 'truth in states of affairs' because we understand 'necessarily'; not *vice versa*.

2.3. Further, if what is 'really meant' by 'Grass is green' is 'The actual world is grass-being-green-ish', what this in turn 'really means' will have to be 'The actual world is the-actual-world-being-grass-is-green-ish-ish', and so on *ad infinitum*. Certainly to be the case is to be the case in the actual world; but this equivalence does not give the meaning of being the case, but rather of the phrase 'in the actual world'. To be the case in a possible world is to be possibly the case; to be the case in an imagined world is to be imagined to be the case; to be the case in a former world is to have been the case; and to be the case in the actual world is just — to be the case. (Cf. Ramsey on truth.) Here again it is the straightforwardly modal calculus of 1.1, without the *x*'s, that brings out the structure of the facts, and when the *x*'s of the calculus of 1.2 are brought into modal logic as names of possible worlds, they are merely obfuscating. To use a distinction I once heard Quine insisting upon, what we have in the calculus of 1.2 may be a *model* for modal logic, but it is not an *interpretation* of the modal words.

3.1. Another special use of the calculus of 1.2 (the uniform monadic first-order predicate calculus) would be this: I don't let my *x*'s stand for any old individuals, but neither do I let them stand for such recondite individuals as possible worlds; I let them stand, rather, for individuals of a *fairly* special sort, namely times or instants, the privileged individual represented by the one variable x when free being the *present* instant. Similarly I don't let my p's, q's, etc. stand for any old predicates but for predicates of the type '— has the proposition p true at it'; i.e. px means 'The instant x has the proposition p true at it', i.e. 'It is the case at the instant x that p'. The 'propositions' here referred to will of course have to be of such a sort that at least some of them will be the case at some instants but not at others; i.e. propositions like 'Nixon is President' and 'Johnson was President'. I call these 'tensed' propositions. The plain px will now mean 'The present instant is a p-ish one', i.e. 'It is the case at the present instant that p', i.e. 'It is now the case that p', i.e. 'It *is* the case that p'. $Lxpx$ will mean 'Every instant is a p-ish one', i.e. 'It is the case at every instant that p', i.e. 'It is always the case that p', or for short, 'p always'. $Mxpx$ will mean 'At least one instant is a p-ish one', i.e. 'It is the case at some instant that p', i.e. 'p at some time'.

3.2. If, having given the x's this interpretation, we now drop them, again obtaining the calculus of 1.1, i.e. S5, this latter will now be read as a fragment of the logic of tensed propositions. In this, what 'is the case at the present instant' is simply what *is* the case, as contrasted with what has been or will be the case, what 'is the case at some instant' is what is or has been or will be the case, this complex of tenses being abridged to the simple modifier M. In a richer logic of tensed propositions we would of course have symbols for the separate components of this complex, say F for 'It will be that' and P for 'It has been that', and this richer tense-logic could also be presented, formally, as a result of deletions from a first-order theory; but it would have to be from a richer theory than the uniform monadic first-order predicate calculus. I shall not go into the details of this, except to say that this richer first-order theory could still be one in which only a single *free* variable occurs, though others would have to occur as bound.

3.3 Given this interpretation of the calculi of 1.1 and 1.2, which is now the more metaphysically illuminating of the two—the calculus with x's, or the one without? Here too I want to say 'The one without'; but here, perhaps, I am swimming against a stronger stream. I think I know what I mean when I say 'Some things are not the case but merely have been or will be the case', better than I know what I mean when I say 'Some things are not the case at this instant but only at earlier or later ones', and certainly better than I know what I mean when I say 'This instant is not *p*-ish but some earlier or later ones are'. And I understand '*p* for ever' better than '*p* at all instants'. I take instants, in short, with the same grain of salt as I take possible worlds.

4.1. Logicians have tended to welcome the presentation of modal logic as an artificially truncated bit of predicate calculus because we know all about predicate calculus, or at all events know an enormous lot about it, whereas modality is a comparatively obscure and unfamiliar field. And even philosophically, it might be said, it is in general pretty clear what is going on in predicate calculus, but not very clear what is going on in modal logic, or even in tense logic. It is not as simple as this. What we can do with first-order predicate logic *in toto* is indeed plain enough; but its uniform monadic fragment? Formally, this fragment is no doubt of some interest; for example, unlike the full first-order predicate calculus, it is decidable. But what is its *philosophical* interest? That question, I think, partly boils down to this one: What would a *philosophically* privileged individual be? And to this question, modal logic and tense logic possibly provide an answer. It is not that modal logic or tense logic is an artificially truncated uniform monadic first-order predicate calculus; the latter, rather, is an artificially expanded modal logic or tense logic. We treat the propositions of modal logic or tense logic as if they were predicates; invent subjects for them, calling these 'possible worlds' or 'times'; make the strong modal or tense operator a universal quantifier, and the weak one a particular quantifier, binding variables standing for these quasi-individuals; the 'privileged' quasi-individual represented by the free variable being a kind of zero attachment to its proposition, so that 'It is the case with the privileged quasi-individual (i.e. in the actual world, at the present time) that *p*' amounts to the plain *p*. If modal logic or tense logic is 'blown up' in this way, what we get is uniform monadic first-order predicate calculus; that is what this calculus 'really is'.

4.2. Or let us say more circumspectly, those are two things that uniform monadic predicate calculus *could* be. There is also a less recondite possibility. In any language with the normal stock of personal pronouns, there is at least one individual with 'privileges' that are perhaps of philosophical interest, namely the speaker. In other words, what the free x's in the calculus of 1.2 might stand for is simply: 'I'. It is arguable, epistemologically, that this is in fact the only individual that any speaker is ever in a position to refer to directly or specifically; all the rest must figure both in his language and his thought as mere 'something-or-others', variously qualified. Or to put it another way, it is arguable that 'I' is the only 'logical proper name' in the Russellian sense of that phrase. If this is so, the use of a predicate calculus with only one name variable occurring free has some philosophical interest and justification.

4.3. Is 'I' in fact the only Russellian individual name? What's wrong with 'this', for example? One argument against 'this', which I owe to Dr. A. J. P. Kenny, could be stated as follows: If 'this' is a logically proper name, then when we say 'This is red', for example, we are simply ascribing redness to a certain individual object without otherwise describing that object in any way. Hence if we say 'This is red' on two different occasions, indicating the same object both times, we are *saying the same thing* on the two occasions. But we may indicate the same object on two occasions without knowing we are doing so; hence we may say the same thing on two occasions without knowing we are doing so, and this seems a little strange, certainly if we know what we are saying on each occasion and remember the first occasion on the second one. Dr. Kenny drew from this argument the conclusion that if there are any logical proper names in the Russellian sense, they can only name momentary objects. As I don't believe in momentary objects, this conclusion, if established by Kenny's argument, would in my view entail that there are no logical proper names in the Russellian sense. However, the indicator 'I' does not seem to me open to Kenny's objection; if I say 'I am in pain' on two different occasions, I cannot but know what object I am indicating, and cannot but know that I am indicating the same object on the two occasions, and therefore (since each time what I ascribe to the object is being in pain) that I am saying the same thing on the two occasions. But so far as I can see, 'I' is the only indicator that is not open to Dr. Kenny's objection.

4.4. So the uniform monadic first-order predicate calculus makes some sense, or has some *rationale*, if we take its one individual variable when free to indicate the speaker, i.e. to do duty for the first person pronoun. But we have seen that where only a single individual variable is used, that variable can be dropped from the calculus without any loss of intelligibility or necessary alteration in its meaning. So we have now a new interpretation of the calculus without the *x*'s, i.e. of S5. We can read the axiom *CLpp* as 'If everything is *p* then I am *p*'; *CpMp* as 'If I am *p* then something is *p*'; and the more complicated S5 law

CLCpLqCMpq,

being a contraction of

CLxCpxLxqxCMxpxqx,

will mean 'If everything is such that if it is *p* then everything is *q*, then if something is *p*, I am *q*', i.e. 'If everything could truly say "If I am *p* then everything is *q*", then if something could truly say "I am *p*", then I can truly say, "I am *q*" '.

4.5. Is this last move—dropping the *x*'s from the calculus of 1.2 when this is 'egocentrically' interpreted—just a formal dodge, or has it also some philosophical significance? I am inclined to think it is just a formal dodge; but it *could* have some philosophical significance, as there are philosophers who are as sceptical about pure egos, and perhaps about individuals generally, as I am about possible worlds and about temporal instants, and such philosophers might well argue that if 'I' is the only logically proper name then indeed there are no logically proper names, since 'I' is not in reality a name of any sort, and is an entirely dispensable expression. I do not wish to develop this point of view in any great detail, but one might sketch it thus: Statements like 'I am in pain', 'I am tall', might be less misleadingly expressed as 'It is hurting', 'The ground is a long way off'. In the former, the 'it' is no more a genuine subject than in

'It is raining', and in fact elementary propositions do not need subjects; the assumption that they do reflects an erroneous 'thing-quality' or 'substance-accident' metaphysic. Of course it is one thing for me to be hurt and another for you to be hurt—your pain is not my pain. But it is a 'remote' pain of mine in the sense in which a former pain or a possible pain is a remote pain, though the 'removal' is in a different direction. Your pain is not what I mean by 'It hurts', but 'It hurts you' is a modality of 'It hurts' in the sense in which 'It could hurt' and 'It did hurt' are modalities of 'It hurts'.

4.6. Of course your pain *is* what *you* mean by 'It hurts', just as my former pain is what I formerly meant, and my possible pain is what I might have meant, by 'It hurts'. But I cannot speak truly for you, any more than I can speak truly today as if it were yesterday. And in this 'I'-less yet egocentric language, *who* is speaking is determined by what is said. Each person is identified by a set of propositions which describe the world as it is from his point of view, i.e. by the set of propositions which are true when said by him; by any single proposition which is true *only* when said by him.

5.1. Whether or not we collapse our language about individuals from that of 1.2 to that of 1.1, and whether or not we regard such a collapsing as a metaphysically significant move, it seems true that each speaker or thinker is himself the only individual he can identify with any assurance, and the rest are, as I said before, mere 'Somethings' variously qualified. It is these 'qualifications' which I now want to examine. We may be aware, for example, not merely that something, but that something that is p, is q. How to symbolise this is obvious; using the x's, we render it as $MxKpxqx$, 'For some x, x is both p and q'. Without the x's, we have $MKpq$, 'Somewhere p and q together' which we may compare with 'Possibly p and q together', i.e. 'p and q are compatible', and 'At some time p and q together', i.e. 'p and q simultaneously'.

5.2. We can also say, still more specifically, '*The* thing which is p is q', and 'q at *the* time at which p', and perhaps also 'q in *the* possible state of affairs in which p'. This, however, is more than can be symbolised in the uniform monadic predicate calculus; it needs at least two individual variables and the dyadic predicate of identity, *I*. Given these, we can render 'The thing which is *p* is *q*' as the conjunction of 'Something which is *p* is *q*' and 'There is exactly one thing which is *p*'. The latter comes out as *MxLyEpyIyx*, 'For some *x*, for all *y*, *y* is *p* if and only if it is identical with *x*', i.e. 'There is an individual with which everything that is *p* is identical'. No additional *free* variables, however, are needed for this, and we would not even need extra bound ones if we introduced 'For exactly one *x*, —' as a special undefined mode of binding variables. We could write it as *Qx*, so that *Qxpx* would mean 'Exactly one thing is *p*', and '*The* thing which is *p* is *q*' could be written as *KQxpxLxCpxqx*, 'Exactly one thing is *p*, and everything that is *p* is *q*'. If we do enrich the uniform monadic first-order predicate calculus with this special quantifier *Q*, it will have to have its own special postulates, and it is now known what these must be, namely a set which I gave—without *x*'s—in the last chapter. My conjecture that these were sufficient has since been verified, independently, by D. Kaplan, K. Fine and R. A. Bull.

5.3. We could also symbolise 'The (one and only) thing that is *p* is *q*' directly, say as *Txpxqx*. Here *Tx* is a quantifier which forms a sentence not from one but from two following sentences. $Qx\alpha$ could then be defined as $Tx\alpha\alpha$, 'Exactly one thing satisfies α' being logically equivalent to 'The one and only thing that satisfies α satisfies α'. This is a bit like developing ordinary quantification theory with formal implication rather than the universal or existential quantifier as a primitive. We could use, say $Fx\alpha\beta$ for 'For all *x*, if α then β', and define $Lx\alpha$ as $FxFx\alpha\alpha\alpha$, 'For all *x*, α' being equivalent to 'For all *x*, α is implied by "For all *x*, if α then α" '. Dropping the *x*'s, S5 necessity can be similarly defined in terms of strict implication, $L\alpha$ as $FF\alpha\alpha\alpha$'[1] That 'the' might be represented by an analogous 'dyadic quantifier' was already suggested;[2] to this suggestion we may now add that within the single-name-variable fragment of predicate calculus that variable can be dropped from *Q* or *T* as well as from *L* or *F*, and we have the same varied pos-

[1] Systems of this sort are discussed in C. A. Meredith and A. N. Prior, 'Investigation into Implicational S5', *Zeitschrift für Mathematische Logik*, vol. 10, no. 3 (1964).

[2] A. N. Prior, 'Is the Concept of Referential Opacity Really Necessary?', *Acta Philosophica Fenica*, fasc. 16 (1963).

sibilities of interpretation. That is, we can read Tpq either as 'The thing that is p is q', or as 'In the state of affairs in which p, it is the case that q', or as 'At the instant at which p, it is the case that q'; the word 'the' being understood in each case as conveying uniqueness, i.e. Tpq is false if nothing or more than one thing is p, or if p is the case in no state of affairs or in several, or at no instant or at several.

6.1. For the function $T\alpha\beta$ I shall now set up postulates. I shall omit x's, as they are only in the way, and therefore subjoin the postulates to any sufficient set for the modal system S5. And as our calculus is purely propositional, it will be convenient to use axioms with substitution for variables, rather than schemata. The axioms are as follows:

T1. $CTpqNLNp$ i.e. $CTpqMp$
T2. $CTpqATprTpNr$
T3. $CTpqLTpq$
T4. $CTpqCpq$
T5. $CLEpqCTppTqp$.

T1 asserts that if the thing that is p is q then something is p; or that if it is the case in the state of affairs at which p that q, then possibly p; or that if it is the case at the instant at which p that q, then at some time p. Its independence may be shown by reading $T\alpha\beta$ as $LN\alpha$. This turns T2–5, but not T1, into theses of S5. T2 asserts that if the thing that is p is q (and so, if there is such a thing as the thing that is p), then either the thing that is p is r, or the thing that is p is not r; or that if anything is the case in the one state of affairs in which p, then either it is the case in that state of affairs that r or it is the case in it that not r; or similarly with instants. The independence of T2 may be shown by reading T as LK. T3 asserts that if the thing that is p is q then it is true of everything that the thing that is p is q; and analogously. If it is understood that variables are 'modalised' by falling within the scope of a T as well as by falling within the scope of an L, T3 may be dropped as an axiom and derived by applying RL of 1.1 to $CTpqTpq$. Its independence, otherwise, may be shown by reading T as K (which does not modalise). That T3 remains independent even when T4 is strengthened to *CTpqLCpq* (which is proved below, from the present basis, as T7), may be shown by reading *Tαβ* as *KαTαβ*, the interpreting *T* being normal. T4 asserts that if the thing that is *p* is *q*, then if I am *p* I am *q*; or that if it is the case in the world in which *p* that *q*, then if it is actually the case that *p* it is actually the case that *q*; or that if it is the case at the time at which *p* that *q*, then if it is the case now that *p* it is the case now that *q*. Its independence may be shown by reading *T* as *MK*. T5 asserts that if exactly the same objects as are *p* are *q*, then if the thing that is *p* is *p* so is the thing that is *q*; or, etc. Its independence may be shown if we use *I* for perfect identity, *k* for some constant proposition, and read *Tαβ* as if it were *KTαβIαk*. For consistency, read *T* as *K* and *Lα* as α; all the postulates (including the S5 ones) then become propositional-calculus theses. My remaining sections will be occupied with deductions from these postulates (and, at the end, from one other).

6.2. I begin by showing the sufficiency of these postulates by deriving from them the postulates for *Q*, whose sufficiency is known. I shall assume laws and rules of S5 without proof.

T6. *CLTpqLCpq* (T4; $\vdash C\alpha\beta \rightarrow \vdash CL\alpha L\beta$)
T7. *CTpqLCpq* (T3, T6)
T8. *CTpqALCprLCpNr* (T2, T7)
T9. *NTpNp* (T7; *LCCpNpLNp*; T1)
T10. *CTpqTpp* (T2 *r*/*p*; T9)
T11. *CLEpqCTppTqq* (T5, T10)
T12. *CLCpqCTppTpq*.

Proof: *C*(1)*LCpq*
C(2)*Tpp*
K(3)*CTpNqLNp* (T7, 1)
K(4)*NTpNq* (T1, 3)
(5)*Tpq* (2; T2; 4).

Defining *Qα* as *Tαα*, we now have

T13. *CQpMp* (T1 *q*/*p*; Df. *Q*)
T14. *CQpALCpqLCpNq* (T8 *q*/*p*, *r*/*q*; Df. *Q*)

T15. *CQpLQp* (T3 *p*/*q*; Df. *Q*)
T16. *CLEpqCQpQq* (T11, Df. *Q*)
T17. *ETpqKQpLCpq* (T10, Df. *Q*; T7; R12, Df. *Q*).

Here T13–16 are the axioms Q 1–4 of the last chapter, and T17 corresponds to the definition of *T*αβ, in a *Q*-primitive system, as *KQ*α*LC*αβ.

6.3. Some further theses that are of some interest are the following:

T18. *CLCpqCTrpTrq*
Proof: *C*(1)*LCpq*
C(2)*Trp*
K(3)*LCrp* (2, T7)
K(4)*LCrq* (1, 3)
K(5)*Trr* (2, T10)
(6)*Trq* (4, 5, T12).

T19. *CLEpqCTprTqr*
C(1)*LEpq*
C(2)*Tpr*
K(3)*Tpp* (2, T10)
K(4)*Tqp* (1, 3, T5)
K(5)*Tqq* (4, T10)
K(6)*LCqr* (2, T7; 1)
(7)*Tqr* (5, 6, T12).

T20. *CLEpqETprTqr* (T19, *ELEpqLEqp*)

T21. *CLEpqETrpTrq* (T18; T18 *p*/*q*, *q*/*p*).

T20 and T21 show that necessarily (universally, permanently) equivalent formulae may replace one another both as first arguments and as second arguments of *T*. Note also that T5 is a substitution instance (*r*/*p*) of T19, which could therefore replace it as an axiom. On the other hand T19 does not *have to* replace T5 as an axiom; given the rest of the system, the apparently weaker T5 yields the apparently stronger T19, as well as *vice versa*. In this T5 is like certain formulae discussed by Sobociński,[1] and by Łukasiewicz.[2]

It would have been possible (though less elegant) to replace axiom T1 by the pair T9 (*NTpNp*) and T12 (*CLCpqCTppTpq*). When these are added to T2–5, independence of T9 may be shown by reading *Tαβ* and *LNα*; and of T12 by reading *T* as *E* and *Lα* as *α*. (The latter as well as the former could be used, with the original axiomatisation, to show the independence of T1.) T10 having been proved from T2 and T9 as in the last section, T1 comes from T9 and T12 by this proof *ad absurdum*:

C(1)*Tpq*
C(2)*LNp*
K(3)*LCpNp* (2)
K(4)*Tpp* (1, T10)
(5)*TpNp* (3, 4, T12),

contradicting T9. With this axiomatisation, T12 is related to T18 as T5 to T19.

7.1. Suppose we use the special variables *a*, *b*, *c*, etc. for those propositions which are true of exactly one individual (in exactly one world; at exactly one time). We may substitute these for the more general propositional variables *p*, *q*, *r*, etc. in theses but not *vice versa*; and in general if *a* and *b* are propositions of this special sort, *Na*, *Cab*, *La* and *Tab* will not be, so that such formulae will not be substitutable in theses for the special variables, though they will be substitutable for ordinary propositional variables. For the special variables we now lay down a single special axiom, namely

T22. *Taa* (i.e. *Qa*)

Adding this to T1–5, the following theses are now provable:

T23. *NLNa* (T22, T1)
T24. *CTaNpCTapLNa* (T7, *CLCpqCLCpNqLNp*)
T25. *CTaNpNTap* (T24, T23)

[1] B. Sobociński, 'An Axiom-system for (K, N)-Propositional Calculus related to Simons' Axiomatisation of S3', *Notre Dame Journal of Formal Logic*, vol. 3, no. 3 (July 1962), p. 208.

[2] J. Łukasiewicz, *Z zagadnien logiki i filozofii*, Warsaw 1961.

T26. *CNTapTaNp* (T22, T2)
T27. *CTpCqrCTpqTpr*
Proof: *C*(1)*TpCqr*
C(2)*Tpq*
K(3)*LCpCqr* (1, T7)
K(4)*LCpq* (2, T7)
K(5)*LCpr* (3, 4)
K(6)*Tpp* (2, T10)
(7)*Tpr* (5, 6, T12).

T28. *CTpKqrTpq* (T18)
T29. *CTpKqrTpr* (T18)
T30. *CTpKqrKTpqTpr* (T28, T29)
T31. *CTaKpNqKTapNTaq* (T30, T25)
T32. *CNKTapNTaqNTaKpNq* (T31, transp.)
T33. *CCTapTaqNTaKpNq* (T32, *NKpNq* = *Cpq*)
T34. *CCTapTaqTaKpNq* (T33, T26)
T35. *CCTapTaqTaCpq* (T34, *NKpNq* = *Cpq*, T21)
T36. *ETaNpNTaq* (T25, T26)
T37. *ETaCpqCTapTaq* (T27, T35).

T36 and T37 are theses which figure prominently in theories about 'truth at an instant', or 'truth in a world'.[1] And the form $Ta\alpha$ may indeed be read as 'It is true in world *a* that α', or 'It is true at the instant *a* that α', if we follow the suggestion made in 2.2 that to be the case in a world or at an instant is to be the case when something else is the case, a given 'world' or 'instant' being now identified with a proposition of a particular sort, a 'state-description'.

7.2. The suggestion of 2.2 that to be true at or in a given state is to be necessarily or permanently implied by that state, is supported by the following deduction:

T39. *CLCpqCLCpNqLNp* (S5)
T40. *CLCapNLCaNp* (T39, T23)
T41. *CLCapNTaNp* (T40, T7)
T42. *CLCapTap* (T41, T26)
T43. *ETapLCap* (T7, T42).

[1] E.g. in *Papers on Time and Tense*, ch. XI.

Note, however, that we cannot prove the more general *ETpqLCpq*. For independence of this, read *T* as *K*, *L*α as α, and let the special variables be restricted to tautologies. All theses of our system will then become tautologies, but *ETpqLCpq* will become *EKpqCpq*, which is not one, being false when *p* is false. And even on the normally intended interpretation, 'Anything that is *p* is *q*' (*LCpq*) will be true, and 'The one and only thing that is *p* is *q*' will be false, if nothing is *p*. However, it is only where there is such a thing as 'the thing that is *p*' that propositions of the form 'The thing that is *p* is *q*' are of much interest, and in those cases, as T43 indicates, 'The thing that is *p*' *is* equivalent to 'Anything that is *p*', *T* to *LC*. So dropping the special operator *T* in favour of *LC*, and leaving the individualising to be done by the special variables, is a move with something to be said for it. If we do this, replacing *T* by *LC* throughout our formulae, the axioms T3–5 and T22 will all become theses of S5, but not T1 and T2. These last could be replaced by T23 (*NLNa*) and T26 in the form *ALCapLCaNp*, laid down as special axioms for the *a*'s. Or we could use T36 in the form *ELCaNpNLCap*, or *ENLCaNpLCap*, or *EMKapLCap*, as a single special axiom for the *a*'s. Or more clumsily, we could use T1 and T2 themselves, with *p* replaced throughout by *a*, as special axioms for the *a*'s. Whichever we do, we obtain all of T1–T43, either as they stand (thus T3–7, T10–12, T15–43) or with *p* replaced by *a* (thus the rest), and in some cases when *T* is replaced by *LC* we can replace *a* by one of the unrestricted variables, say *s* (thus T22, T24, and of course T42 and 43). Consistency of the system may be shown by reading *L*α as α and confining *a*'s to tautologies, and if we axiomatise with T23 (*NLNa*) and T26 (*ALCapLCaNp*), independence of the latter may be shown by reading *L* normally and confining *a*'s to tautologies, and of the former by reading *L* normally and confining *a*'s to contradictions.

5. *Tensed Propositions as Predicates*

Prima facie, we may divide propositions, or ostensible propositions, into two sorts. There are those, such as 'Socrates is sitting down' or 'Socrates was sitting down', of which it obviously makes sense to ask '*When* are they true?', though the answer may in some cases be 'Always' or 'Never'. And there are, on the other hand, those of which it does not so obviously make sense to ask this question; one sub-species of these would be exemplified by 'Two and two are four', and another by 'The date of the Battle of Hastings is 1066'. Rescher has suggested that propositions in the first major division (those of which at least some are true at some times and false at others) be called 'chronologically indefinite', while those in the second major division he calls 'chronologically definite'.[1] In what follows, I shall call propositions of the first sort 'tensed', and propositions of the second sort 'untensed'. One view of the relation between the two sorts of propositions is that 'untensed' propositions are improperly so called; they are simply tensed propositions which are so obviously either always or never true that to ask *when* they are true is socially (but not logically) inept. The most plausible alternative to this is to say that tensed propositions are not genuine propositions but predicates of instants ('Socrates is sitting down' really means 'Socrates is sitting down at —'), the instants which form their subjects being unmentioned, but usually understood to be the time of utterance.

Of these two views about the relations between tensed and tenseless propositions, I am myself a determined adherent of the first, i.e. the view that so-called tenseless propositions are simply a sub-class of those tensed propositions which are always or never true. It is not my present purpose, however, to defend this view, but rather to see how far we can take the opposite view, i.e. the view

[1] N. Rescher, 'On the Logic of Chronological Propositions', *Mind*, 75 (1966).

that tensed propositions are not propositions at all, but predicates of instants. I shall take my departure from my paper of 1967 on 'Tense Logic and the Logic of Earlier and Later',[1] hereinafter called TEL, in which I outline a series of formal systems concerned with the notion of a proposition or quasi-proposition p being true at an instant a. In the system which I call System I, and also in a slight enlargement of it which is given no number in TEL but which I shall here call System I+, tensed propositions are a fairly restricted class; in particular, propositions to the effect that such and such a tensed proposition is true at such and such a time are not themselves classed, in this system, as tensed propositions. Nor are propositions to the effect that such and such an instant is earlier than such and such another instant. These restrictions are removed in the system which I call System II. And in System III, instants themselves are treated as tensed propositions of a sort (each instant is identified with a tensed proposition which would ordinarily be said to be true at that instant only). Using Systems I and I+, in which tensed propositions are syntactically separated from propositions about instants, it is plausible to regard these tensed propositions as predicates of the instants at which they are said to be true; but this way of looking at the matter is hardly consonant with the syntax of System II, and still less with that of System III. I therefore describe these systems, in TEL, as embodying increasing 'grades of tense-logical involvement'.

In the present paper I make both formal and philosophical revisions of TEL. In the first place, I make the syntax of the various systems more explicit than it is in TEL. Secondly I reaxiomatise them in a way which I hope will be more illuminating (and which will at two points be more correct) than the axiomatisations of TEL. Finally, I show that the view of tensed propositions as predicates, though I still think it misguided, can be carried further than is admitted in TEL. I do this last by making certain modifications to the more 'tense-logically involved' of my systems, and tracing some of their consequences. In the course of these manœuvres, I hope that some new light may be thrown not only on tense logic but also on predicate calculus.

[1] *Papers on Time and Tense*, ch. XI.

1. *The systems I, I+, II, IIIa, IIIb, IIIc+ and IIIc*

The vocabulary of System I consists of the instant variables a, b, c, etc., the tensed propositional variables p, q, r, etc., and the undefined constants C, N, G, H, I, U, T, and Π. Tensed propositions are defined as follows:

(i) The variables p, q, r, etc., are tensed propositions.
(ii) If ϕ and ψ are tensed propositions, so are $C\phi\psi$, $N\phi$, $G\phi$, and $H\phi$.

Informally, $C\phi\psi$ is to be read as 'If ϕ then ψ', $N\phi$ as 'Not ϕ', $G\phi$ as 'It will always be that ϕ' and $H\phi$ as 'It has always been that ϕ'. Other truth-functions, in particular $A\phi\psi$ ('Either ϕ or ψ'), $K\phi\psi$ ('Both ϕ and ψ') and $E\phi\psi$ ('If and only if ϕ then ψ'), may be defined in the usual ways, and $F\phi$ (for 'It will be that ϕ') and $P\phi$ (for 'It has been that ϕ') as $NGN\phi$ and $NHN\phi$ respectively. Untensed propositions are defined as follows:

(i) If u and v are instant variables, Iuv and Uuv are untensed propositions.
(ii) If u is an instant variable and ϕ a tensed proposition, $Tu\phi$ is an untensed proposition.
(iii) If α and β are untensed propositions and u is an instant variable, $C\alpha\beta$, $N\alpha$ and $\Pi u\alpha$ are untensed propositions.

Informally, Iuv is to be read as 'u is the same instant as v', Uuv as 'u is earlier than v', $Tu\phi$ as 'It is the case at u that ϕ', and $\Pi u\alpha$ as 'For all u, α'. C and N are for 'If' and 'Not', and the other truth-functions definable in terms of them, as before; and $\Sigma u\alpha$, for 'For some u, α', is definable as $N\Pi uN\alpha$.

All the theses of the system are untensed propositions, and the postulates consist of the following axioms subjoined to ordinary propositional calculus, quantification theory, and identity theory:

ET1. $ETaCpqCTapTaq$
ET2. $ETaNpNTap$
UT1. $ETaGp\Pi bCUabTbp$
UT2. $ETaHp\Pi bCUbaTbp$.

Or in words:

ET1. It is true at a that if p then q, if and only if, if it is true at a that p it is true at a that q.
ET2. It is true at a that not p, if and only if it is not true at a that p.
UT1. It is true at a that it will always be the case that p, if and only if it is true at all instants later than a that p.
UT2. It is true at a that it has always been the case that p, if and only if it is true at all instants earlier than a that p.

As observed in TEL, the symbol T is superfluous in this system; we could write $Tu\phi$ as ϕu and regard the axiom as defining complex predicates in terms of propositional complications, thus:

ET1. $(Cpq)a = Cpaqa$ Df.
ET2. $(Np)a = N(pa)$ Df.
UT1. $(Gp)a = \Pi bCUabpb$ Df.
UT2. $(Hp)a = \Pi bCUbapb$ Df.

Or in words:

ET1. 'a is an if-p-then-q-ish instant' = 'If a is p-ish then a is q-ish'.
ET2. 'a is a not-p-ish instant' = 'a is not a p-ish instant'.
ET3. 'a is a p-for-evermore-ish instant' = 'All instants later than a are p-ish'.
ET4. 'a is a p-through-all-the-past-ish instant' = 'All instants earlier than a are p-ish'.

It is also stated, erroneously, in TEL that the equivalence ET1 can be weakened to an implication; in fact this weakening is only possible in System II. The proof of ET1 from the corresponding implication requires the rule that if α is a thesis so is $Ta\alpha$; in System I either the α or the $Ta\alpha$ of this rule will not be an untensed proposition as above defined. However, for System I as now axiomatised we can prove the metatheorem that if the tensed proposition ϕ is tautological in form, $Ta\phi$ is a thesis. (Just use ET1 and ET2 to equate the tensed tautology, preceded by Ta, to an untensed one; e.g. *TaCpp* to *CTapTap*.) And, as mentioned in TEL and elsewhere, the addition of UT1 and 2 enables us to prove, preceded by *Ta*, all theses of Lemmon's minimal tense logic K_t, the postulates of which I need not repeat here.[1]

In System I+, we enlarge the class of tensed propositions by the following additions:

(iii) If one of ϕ and ψ is a tensed proposition and the other untensed, $C\phi\psi$ is a tensed proposition.

(iv) If *u* is an instant variable and ϕ a tensed proposition, $\Pi u\phi$ is a tensed proposition.

And for these new tensed propositions, we add to the axioms of System I the following axiom-schemata:

ET1.1 $ETaC\alpha pC\alpha Tap$
ET1.2. $ETaCp\alpha CTap\alpha$
ET1.3. $ETa\Pi b\phi\Pi bTa\phi$

where ϕ is a tensed and α an untensed proposition. (The necessity for ET1.3 was overlooked in TEL.) These equivalences can also be replaced by definitions, and we can regard this enlargement as simply allowing us to form complex predicates not only from other predicates but from a combination of these with propositions. (This is normal procedure in predicate calculus.) And it may be noted that each of the schemata ET1.1–3 equates something which is ill-formed in System I with something which is well-formed there; thus $TaC\Pi bTbpb$ (e.g. 'It is true at *a* that if Socrates is at all times sitting down then Socrates is sitting down') is equated by ET1.3 with $C\Pi bTbpTap$ ('If it is true at all times that Socrates is sitting down then it is true at *a* that he is').

In System II the entire class of untensed propositions is absorbed into the class of tensed ones, and tensed propositions are admitted as theses. This makes the syntactical enlargements of System I+ redundant, and its postulates ET1.1–3 are also made redundant by the postulates which System II adds to System I, namely

[1] See, e.g., ibid., p. 116.

RET: ⊢*a* if and only if ⊢*Taa*, for *a* not free in *a*.
ET3. *ETaTbpTbp*
ET4. *ETa*Π*bTbp*Π*bTbp*
ET5. *ETaUbcUbc*.

These postulates are equivalent to those given for System II in TEL, and, as there observed, the ET portion of the system is equivalent to a system of Rescher's. We cannot now dispense with the functor *T* in favour of simple predication without using propositions as predicates, i.e. having forms like (*pa*)*b* and ((*pa*)*b*)*c*. However, each of the axioms above equates a formula which is ill-formed in System I with one which is well-formed there, and the rule has a similar function.

Finally, in System III, the class of tensed propositions absorbs even instants, and for instants considered as propositions we have the one further axiom (which makes ET5 redundant).

ET6. *ETabIab*,

asserting that the 'instant' *b* is 'true at' *a* if and only if it is identical with *a*. The identity function is of course taken to have its usual postulates (say *Iaa* and the schema *CIabC*$\alpha\beta$, where β differs from α only in having a *b* in place of an *a*). In TEL, *I* is not thus used as a primitive, though the possibility of doing so is mentioned. System III seems as far removed as possible from the representation of tensed propositions as predicates of instants—instants as objects have just disappeared. But (1) ET6 is like earlier extensions of System I in equating a formula which is ill-formed in that system with one which is well-formed there, though the *I* of *Iab* must be re-interpreted as some sort of connective when *a* and *b* are read as propositions; and (2) a very slight modification to System III will make it much easier to interpret its tensed propositions as predicates.

The modification I have in mind consists in dropping the identification of an instant with a tensed proposition true at that instant only, and introducing a functor *S* which forms that proposition from that instant. The tensed proposition *Sa* may be read as 'It is *a*', in the sense of 'it is' in which we say 'It is 5 o'clock', 'It is Thursday', 'It is the 5th of June', etc. And instead of adding ET6 to System II, we add

S1. *ETaSbIab*,

'That *it is b* is the case at *a* if and only if *a* and *b* are the same instant'. Moreover, the form *Sa* and the axiom S1 could be added, not to System II, but to System I+ or even System I without doing violence to the syntax of those systems; though if the form *Sa* is introduced into System I as a tensed proposition, it is natural to introduce quantifications of tensed propositions, and with them ET1.3, into that system.

For these various possible new systems I propose to adopt the following names:

(1) 'System IIIa' for System III as above presented, i.e. System II with instants counted as tensed propositions and ET6 added to the postulates.
(2) 'System IIIb' for System II with the following added to the definition of a tensed proposition: 'If *u* is an instant-variable, *Su* is a tensed proposition'; and with S1 added to the postulates.
(3) 'System IIIc+' for System I+ with the same additions to the definition of 'tensed proposition' and to the postulates.
(4) 'System IIIc' for System I with the following added to the definition of a tensed proposition: 'If *u* is an instant-variable and ϕ a tensed proposition, *Su* and $\Pi u\phi$ are tensed propositions'; and with ET1.3 and S1 added to the postulates.

In Systems IIIc+ and IIIc, it may be noted, as in Systems I+ and I, the symbol *T* can be dispensed with in favour of predication, and S1 replaced by the definition

$(Sb)a = Iab$. Df.

2. *Quasi-modal fragments of the systems*

Where ϕ does not contain free *a*, we may abridge $\Pi aTa\phi$ ('It is true at all times that ϕ') to $L\phi$, and it is pointed out in TEL that for an *L* thus defined we may prove in System II the Lewis modal system S5, and also *CLpGp* and *CLpHp*. What is not investigated in TEL is the result of introducing this definition of *L* into Systems I+ and I. The most important point to note about this is that in these systems *L* is a 'heterogeneous' functor, forming untensed propositions from tensed ones. *CLCpqCLpLq*, *CLCpqCLCqrLCpr*, and (with *M* for *NLN*) *CLCpqCMq*, *CKLpMqMKpq* and *CLpMp* are examples of formulae which remain well-formed, untensed, and theses in these systems; but *CLpp*, *CLpGp*, and *CLpHp* are ill-formed in System I (being implications with untensed antecedents and tensed consequents), while in System I+ they are tensed and therefore not theses. However, *LCLpp*, *LCLpGp*, and *LCLpHp* are theses of System I+, and *CLpLGp* and *CLpLHp* even of System I.

The fragments of Systems I and I+ which use only *C*, *N*, *L*, and tensed propositional variables may be regarded as notational variants of certain forms of predicate calculus without name-variables developed by G. E. Hughes and D. G. Londey. In their published textbook,[1] Hughes and Londey give a calculus in which predicates consist of predicate-letters *f*, *g*, *h*, etc., and truth-functional complexes of these, atomic propositions are formed by prefixing *U* or *E* (in effect universal and existential quantifiers) to these, and molecular propositions in the usual truth-functional ways. As postulates we have all tautologies formulable in the language, all tautological truth-functional complexes of predicates with the whole preceded by *U*, and the following axioms (for use with predicate-substitution and detachment):

A1. $U(f \supset g) \supset (Uf \supset Ug)$
A2. $Uf \supset Ef$.

Without A2 (which expresses the non-emptiness of the universe), this system would appear to be equivalent to an earlier one with the same syntax, namely von Wright's 'Quantified Logic of Properties'.[2] Von Wright also has the idea of classifying the operators *U* and *E* as 'modalities' ('existential' modalities).[3] At all events, a mere re-writing of the Hughes-Londey postulates in our symbolism will give a basis for the *C–N–L–p* portion of System I, namely: All tautologies formulable in the language, all tautological

[1] G. E. Hughes and D. G. Londey, *Elements of Formal Logic* (New York 1965).
[2] G. H. von Wright, *Logical Studies* (London 1957), second paper ('On the Idea of Logical Truth(I)').
[3] G. H. von Wright, *An Essay in Modal Logic* (Amsterdam 1951), ch. II.

complexes of tensed propositions with the whole preceded by *L*, and the axioms *CLCpqCLpLq* and *CLpMp*.

Hughes has also axiomatised[1] an extension of the above predicate calculus in which propositional as well as predicate variables are used in the formation of complex predicates (an implication of which one component is a predicate counts as a predicate). The axioms are all tautologies, all tautologously formed predicates (including those with propositional components) preceded by *U*, and the axioms

A1.1. $U(p \supset f) \supset (p \supset Uf)$
A1.2. $U(f \supset g) \supset U(Uf \supset g)$
A2. $Uf \supset Ef$.

This gives for the *C–N–L–p* fragment of System I+: all tautologies, all tautologously formed tensed propositions preceded as a whole by *L*, the axiom schema *CLCαpCαLp* (where α is untensed) and the axioms *CLCpqLCLpq* and *CLpMp*. (*LCLpp* is obtained from *LCpp* and the first axiom, and *CLCpqCLpLq* from the first axiom and the schema.)

Nested modalities like *LLp* and *LMp* are of course ill-formed in both these systems, but they are not simply modal logics with such nestings ruled out (of the sort studied, e.g. by J. L. Pollock,[2] and earlier but more sketchily by myself),[3] since even *CLpp* is not propositional enough to figure as a thesis.

If we add the tense-operators *G* and *H*, the resulting enlarged fragments of both systems will contain all theses of the tense-logic K_t preceded as a whole by *L*, and may be axiomatised by adding to the above postulates, besides these *L*-preceded K_t theses, in System I *CLpLGp* and *CLpLHp*, and in System I+ *LCLpGp* and *LCLpHp* (from which *CLpLGp* and *CLpLHp* are obtainable by the schema *CLCαpCαLp*, with *Lp* for α and *Gp* and *Hp* for *p*). One way of obtaining the K_t theses preceded by *L*, in either system, would be to lay down a sufficient set of K_t axioms each preceded by *L*, and then derive the theorems (a) by using substitution, *CLCpqCLpLq* and detachment, and (b) where K_t theses are

[1] See A. N. Prior, critical notice of Hughes and Londey's *Elements of Formal Logic*, in *Australasian Journal of Philosophy*, 44 (1966).
[2] John L. Pollock, 'Basic Modal Logic', *Journal of Symbolic Logic*, 32 (1967).
[3] *Time and Modality* (Oxford 1957), Appendix C.

obtained by using the rules to infer $\vdash G\phi$ and $\vdash H\phi$ from $\vdash\phi$, to obtain the corresponding L-preceded theses by $CLpLGp$ and $CLpLHp$. (Tautological tensed formulae preceded by L could be similarly obtained by laying down L-preceded tensed propositional-calculus axioms and using substitution, $CLCpqCLpLq$ and detachment.)

There is not much advantage in using the richer syntax of System I$^+$, but there is a little. Suppose, for example, we wish to add to either of the systems the postulate that there is a unique time-series, and that a linear one. Using instant variables, we can express this in the *full* vocabularies of both systems by $AAUabUbaIab$, asserting of any instants a and b that a is either earlier than or later than or identical with b; or by $CTaKKpHpGpLp$, asserting that if it is true at any instant a that it is and always has been and always will be the case that p, then p is true at all times. But in the vocabulary of System I$^+$ we can express this assumption in pure L–C–N–G–H–p, by $LCKKpHpGpLp$.

In both systems, G and H are 'extensional' functions of their arguments, in the sense that we have $CLEpqLEGpGq$ and $CLEpqLEHpHq$. If we read these arguments as predicates, this means that G and H are extensional functions in the standard sense, since these laws just mean $C\Pi aEpaqa\Pi aE(Gp)a(Gq)a$ and $C\Pi aEpaqa\Pi aE(Hp)a(Hq)a$.

Considered as bits of predicate calculus without name-variables, the extensions with G and H are of some interest. The Hughes-Londey pedagogical programme of delaying the introduction of name-variables as long as possible begins to run into difficulties when we move from one-place to two-place predicates (how, e.g., without using name-variables, do we distinguish between $\Pi a\Sigma bUab$ and $\Pi b\Sigma aUab$?), and in fact they make no attempt to carry it that far; but here *some* of the logic of the two-place predicate U is brought in via a pair of operators forming one-place predicates from other one-place predicates—in particular, forming '— has only p-ish instants following' and '— has only p-ish instants preceding' from the predicate-of-instants '— is p-ish'. The procedure may be generalized by replacing any relation R by operators, on a predicate f, forming '— is R only to what is f', and '— is R'd only by what is f'.[1] The L–C–N–G–H–p logics we have been consider-

[1] For beginnings of such generalisations, see *Papers on Time and Tense*, ch. XII, and Chapter 2 above.

ing give a general calculus for such pairs of operators, and some well-known extensions of this logic enable us to state, within this type of calculus, certain properties of the 'hidden' relation *U*. For example, it is known that if we add to System I the postulate that *U* is transitive, we can prove not just the theses of K_t preceded by *Ta* but, preceded by *Ta*, the theses of $K_t + CGpGGp$; and in the present fragment of System I, the transitivity of the unmentioned *U* may be expressed by adding the axiom *LCGpGGp*. And the replacement of *U* by *G* and *H* can be taken still further in the Systems III.

3. *The tense-logical construction of the earlier-later logic*

In what I am now calling System IIIa, it was shown in TEL that we may prove *ETapLCap*, and that the function *LE* satisfies the normal postulates for identity within the system. This suggests a re-axiomatisation with *L* as primitive and *T* and *I* defined as *LC* and *LE* respectively. With *L*, *C*, and *N* primitive, and the definitions just given, we obtain a system equivalent to the *T–I* portion of IIIa if we add to S5 for *L* three special postulates for 'instant-propositions', namely *Ma*, *ALCapLCaNp* and *Σaa*.

Further, in System IIIa, with UT1 and 2 included, we can prove *EUabTaFb* and *EUabTbPa* (*a* is earlier than *b* if and only if it is true at *a* that it will be that *b*, and if and only if it is true at *b* that it has been that *a*). This suggests dropping *U* as a primitive and defining *Uab* either as *TaFb* (*LCaFb*) or as *TbPa* (*LCbPa*). If we do this, and add K_t for *G* and *H*, with the mixing postulates *CLpGp* and *CLpHp*, to the *L–C–N* postulates just mentioned, or to the *T–I* postulates to which they are equivalent, UT1 and 2 become provable. Thus the elementary logic of the earlier-later relation *U* can be developed from the logic of the tenses *G* and *H*, given the above postulates for *L* or for *T* and *I*. Moreover, those conditions on *U* which have 'reflections' in tense-logic can be proved from those reflections; and even those which have none can be reformulated in terms of *G*, *H*, and *L*.

These manœuvres, however, presuppose both the identification of instants with instantaneous tensed propositions (the peculiarity of System IIIa) and the identification of *T*, *U*, and *I* functions with omnitemporal tensed propositions (a feature of System II also). The question which we may now ask is: How far can these manœuvres be reproduced in systems such as IIIc+ and c, which do not have these peculiarities, and in which tensed propositions are in effect treated as predicates of instants (just as they are in I+ and I)?

The fact that the postulates of the richer systems can be given as simply equating new forms with old is an encouraging sign for this enterprise, and it turns out that practically everything required can be carried out even in the least 'tense-logically involved' System IIIc. We can prove, even in System IIIc, *ETapLCSap*, *EUabTaFSb*, *EUabTbPSa*, *MSa*, *ALCSapLCSaNp*, and *LΣaSa* (the plain *ΣaSa*, being tensed, is not suitable for a thesis in System IIIc). We can also prove, even in IIIc, the 'Barcan' schema *EΠbLϕLΠbϕ*; an important point because in System II (on which Systems IIIa and IIIb are based) this schema can be represented as a by-product of quantification theory and S5 for *L*, but it is not provable from quantification theory and the *L–C–N–p* fragment of System I (on which IIIc is based).

We can now rebuild IIIc in various ways. In the least radical way, we may dispense with *T* in favour of predication, i.e. replace *Tap* by *pa*, and (1) introduce into predicate calculus with identity the definitions (*Cpq*)*a* = *Cpaqa*, (*Np*)*a* = *Npa*, (*Sb*)*a* = *Iab*; and then (2) instead of adding a primitive *U* (with no axioms) and defining (*Gp*)*a* and (*Hp*)*a* as *ΠbCUabpb* and *ΠbCUbapb*, we could define *Uab* as (*FSb*)*a* or (*PSa*)*b*, and for the undefined *G* and *H* lay down the axioms for K_t, each predicated of an arbitrary instant *a*, and the further axioms *CΠbpb*(*Gp*)*a* and *CΠbTb*(*Hb*)*a* (cf. *CLpGp* and *CLpHp* in System II and its extensions); equivalences corresponding to the definitions of *G* and *H* are then provable. We gain very little, of course, by this; the *U*-primitive version of System IIIc is obviously more compact; the reshuffling merely helps to show the sufficiency of K_t for a tense logic presupposing no special conditions on *U*.

It is possible, however, to proceed quite differently, and more interestingly. We may present the System IIIc as indeed a disguised predicate calculus, but of an extended Hughes-Londey sort; that is, if we may so speak, a predicate calculus without predication. We shall have, indeed, instant variables and quantifiers binding them (our system could not be equated with IIIc without them), but these instant-variables will not figure as subjects of predication—

we shall not represent Tap as a variant of pa; but differently. Instant-variables will occur only as constituents of predicates (or tensed propositions) of the form Su, and Tap will be defined as $LCSap$, i.e. in effect, as the formal implication of the predicate p by the predicate Sa. Nor will S be defined by equating $(Sb)a$ with Iab; on the contrary, S will be one of our primitives and Iab defined as $LESaSb$, i.e. as the formal equivalence of the uniquely instantiated predicates Sa and Sb. L will be primitive, and the only undefined way of constructing a genuine or untensed proposition from a predicate, i.e. tensed proposition, will be by prefixing L to it.

We now build up the postulates of the system in the following steps:

(1) We axiomatise the L–C–N–p fragment of System I as in the last section.

(2) We introduce instant-variables, quantifiers binding them, and S, with the three axioms MSa, $ALCSapLCSaNp$, and $L\Sigma aSa$; the rules

$\Pi 1$: $C\alpha\beta \rightarrow C\Pi u\alpha\beta$
$\Pi 2$: $C\alpha\beta \rightarrow C\alpha\Pi u\beta$, for u not free in α
$L\Pi 1$: $LC\phi\psi \rightarrow LC\Pi u\phi\psi$
$L\Pi 2$: $LC\phi\psi \rightarrow LC\phi\Pi u\psi$, for u not free in ϕ,

(where α, β are untensed propositions and ϕ, ψ tensed); and the schema $C\Pi uL\phi L\Pi u\phi$ (the converse being provable).

(3) We introduce the predicate-formers (tensed-proposition-formers) G and H with postulates as in the last section, i.e. L-preceded K_t, $CLpLGp$ and $CLpLHp$.

(4) We define Tap as $LCSap$, Iab as $LESaSb$, and Uab as $LCSaFSb$ $(TaFSb)$.

All these postulates (including equivalences corresponding to the definitions) are provable in our original version of IIIc; and the latter (including Iaa and the schema $CIabC\alpha\beta$, where α and β are untensed propositions of the system and β differs from α only in a replacement of a by b) is provable from them.

If we construct a similar version of System IIIc+ (with (1) and (3) adjusted as in the last section, and with it understood that in $L\Pi 1$

and LΠ2 *at least one* and possibly but not necessarily both of φ and ψ must be tensed), we can omit the Barcan schema as provable, though we still need *LΣaSa*.

The predicates *Su* function in these systems like Quine's 'Pegasises'. Their unique instantiation is guaranteed by their special axioms. *MSa* asserts in effect that at least one individual (instant) has the property *Sa*, and *ALCSapLCSaNp* that for any property *p* either whatever has *Sa* has *p* or whatever has *Sa* lacks *p*, so that we never have one thing with *Sa* having *p* and another lacking it; i.e. at most one thing has *Sa*.

We could go further than this. To symbolise a uniquely occurring predicate (tensed proposition) as *Sa* still suggests something more like 'is Pegasus' than 'Pegasises' (our first reading of this form was, in fact, something like 'is 5 o'clock', or more precisely 'is 5 o'clock G.M.T. on January 14, 1957'), and we could drop this suggestion of an internal structure by not using the symbol *S* at all and treating the *u*'s (unprefixed) as themselves the uniquely occurring predicates (their unique instantiation being, again, secured by the special axioms). This would give us a system, which we might call IIId, in which the 'tensed propositions' of System I *are* enlarged by treating instants as such propositions, but are enlarged *only* in this way, genuine or untensed propositions (such as *Tab*, *Uab*, *Lp*, and truth-functions and quantifications of these) being still syntactically segregated from tensed ones, and being alone capable of figuring as theses.

4. *Further considerations regarding System IIId*

It is tempting to say that, as far as the removal of dubious presuppositions is concerned, in this last system, IIId, we contrive to get the best of both worlds. Tensed propositions are still resolutely treated as predicates rather than as genuine propositions (so that our 'tense-logical involvement' is still minimal); yet instants do not figure in the systems as named entities, so that we do not seem committed to their 'existence' in any serious sense. We seem, in fact, to have come up with a type of calculus which might very well suit people who say such things as that logic should detach itself from a thing-quality or substance-accident metaphysic, or that things should be regarded as logical constructions out of events rather than *vice versa*. Against this it might be argued that although Hughes-Londey types of predicate calculus do not *mention* individuals they presuppose them, and that if tensed propositions in particular are really predicates we cannot avoid, though we may symbolically gloss over, the question as to what they are predicates *of*. It may also be remarked that even in System IIId we still *quantify* over instant variables, and on some views this would suggest we are not taking their treatment as predicates very seriously.

My own view on these points is that (a) a thing-quality metaphysics is fine, but that (b) instants are not things, and (c) quantifying over instant-variables does not commit us to the view that instants *are* things. But I do not wish to argue for these positions here, but rather to note that formally System IIId, as well as System IIIc, *could* be interpreted by taking the thinghood of instants very seriously indeed, and reading its tensed propositions simply as *class* symbols, denoting *sets* of instants, namely the sets of instants 'at' which we would normally say that each proposition is true. This interpretation is possible because of the extensionality of the functions G and H, considered as functions of predicates. It will be simplest to begin by developing this interpretation of System IIIc.

On this reading $N\phi$ obviously denotes the Boolean complement of the set ϕ, and $K\phi\psi$ the intersection of the sets ϕ and ψ. $L\phi$, which is 'genuinely propositional' in Systems IIIc and d, does not denote a set but a sentence asserting that the set ϕ (i.e. the set of instants at which ϕ) is the universal set (the totality of instants); or in brief, that $\phi = 1$. The set $C\phi\psi$, i.e. $NK\phi N\psi$, is the complement of the intersection of ϕ with the complement of ψ; i.e. that part of the totality of instants which lies outside that part of the set ϕ which does not overlap the set ψ, i.e. the union of the set ψ and the complement of ϕ ($NK\phi N\psi = AN\phi\psi$). $LC\phi\psi$ then asserts that this set $NK\phi N\psi = 1$, i.e. that $K\phi N\psi = 0$, i.e. that no part of the set ϕ fails to overlap the set ψ, i.e. that ϕ is wholly included in ψ (the obvious set-theoretic version of the statement that all instants at which ϕ are instants at which ψ, or that whenever ϕ is true so is ψ). G and H are functions from sets of instants to sets of instants, such that $G1 = H1 = 1$, $GK\phi\psi = KG\phi G\psi$ and $HK\phi\psi = KH\phi H\psi$, i.e. the G-set of the intersection of ϕ and ψ is the intersection of the G-sets of ϕ and ψ, and similarly for H. (These conditions are set-theoretically equivalent to laying down $CL\phi LG\psi$, $CL\phi LH\psi$, $LCGC\phi\psi CG\phi G\psi$, and $LCHC\phi\psi CH\phi$-$H\psi$.) Further, relating the two functions, we have $LCNGNH\phi\phi$ and $LCNHNG\phi\phi$. It can then be shown that there is a relation-in-extension U between the instants in our sets, for which $G\phi$ is the set of instants such that all instants U'd by them are in ϕ, and $H\phi$ is the set of instants such that all instants which are U to them are in ϕ; or in symbols, $G\phi = a{:}\Pi bCUab\phi b$ and $H\phi = a{:}\Pi bCUba\phi b$. For this property will be possessed by the relation between instants a and b which consists in a being in the set $NGNSb$, where Sb is the set of which b is the sole member.

In System IIId, U is not expressible as a relation between instants, since that system does not contain names for instants. However, it has the symbols a, b, c, etc., denoting the corresponding unit sets, and so has the function Uab which consists in the unit set a being included in the set $NGNb$ ($LCaNGNb$), and for U thus defined we can prove that a unit set a is included in $G\phi$ if and only if $\Pi bCUabLC\phi$ and in $H\phi$ if and only if $\Pi bCUabLCb\phi$. This, on this interpretation of the system, is precisely what we prove when we derive UT1 and 2 from its postulates for G and H.

Thus interpreted, System IIId is an 'algebra' in the sense of some recent papers by E. J. Lemmon,[1] and its proofs of theses about U from theses about G and H are simply versions of completeness proofs using such algebras. An extensional predicate calculus without name-variables is after all just a version of set-theory without symbols for members or for membership, and that in turn is simply a Boolean algebra, though if members of sets are to be represented in it by their unit classes it must be an 'atomic' Boolean algebra, i.e. it must contain elements a such that $a \neq 0$ and for any ϕ, if ϕ is included in a then either $\phi = 0$ or $\phi = a$, and such that if any element $\neq 0$ then some atomic element is included in it.[2]

The set-theoretical proofs will of course go through *whatever* the

[1] E. J. Lemmon, 'Algebraic Semantics for Modal Logics', *Journal of Symbolic Logic*, 31 (1966), pp. 46–65, and pp. 191–218.

[2] Cf. Tarski, 'On the Foundations of Boolean Algebra', *Logic, Semantics and Metamathematics* (Oxford 1956), pp. 320–41, esp. 2. Tarski does not have special symbols for atoms, but a definition of atomicity and a postulate for it, as above. This corresponds to variants of our systems in which special instant-variables are dispensed with in favour of the function Qp, for 'p is true at one instant only' (cf. TEL p. 128, and the last section of Chapter 2 above).

sets are supposed to be sets of, e.g. if their members are teacups, $G\phi$ is the set of teacups such that all more expensive teacups than these ones are in the set ϕ, and $H\phi$ is the set of teacups such that all *less* expensive teacups are in ϕ. This, to my mind, means that we do not need to take the 'instants' of the original interpretation terribly seriously after all; for formal completeness proofs we can happily use System IIIc or d and think of teacups, and for metaphysics we can paraphrase the instants away altogether by thinking of IIIc and d as System IIIa with certain not very illuminating types of well-formed formulae (e.g. $TaTbp$) omitted.

6. *Modal Logic and the Logic of Applicability*

The modal system S5 may be axiomatised with N ('Not'), K ('and') and M ('Possibly') as primitives, with C ('If') and E ('If and only if') defined in the usual ways, by subjoining to propositional calculus with substitution and detachment the rule RM to infer $\vdash NMN\alpha$ from $\vdash \alpha$ and the axioms

S1. $CNMNCpqCMpMq$
S2. $CpMp$
S3. $CMNMpNMp$.

The system may be thought of as embodying the notion of truth in possible worlds. If we write Tap for 'It is the case in the world a that p' we may lay down for this T the axioms

T1. $ETaNpNTap$
T2. $ETaKpqKTapTaq$
T3. $ETaMp\Sigma bTbp$,

where T3 states that it is the case in a world a that possibly-p, if and only if in some world it is the case that p. If these axioms are subjoined to first-order predicate calculus we can prove everything in S5 preceded by an arbitrary Ta.

Some time ago I proposed, and gave a matrix for, a modal system Q in which it is assumed that in certain worlds certain propositions simply do not occur, because they are directly about individuals which are absent from those worlds. This system has since been axiomatised by R. A. Bull, K. Segerberg and myself. A convenient axiomatisation to use here is the following: We introduce, beside

K, *N* and *M*, a function *Sp* meaning that *p* occurs, or as I sometimes put it is 'statable', in all worlds, with the two schemata

RS1: ⊢*CSαSp*, for any *p* in α
RS2: ⊢*CSpCSq* . . . *Sα*, where *p*, *q* . . . are all the variables in α.

For *M* we have

M1. *CpMp*
M2. *CMNMpNMp*;

and the rule

RM: If ⊢α then ⊢*NMNα*.

For *M* and *S* together we have the axiom

MS1. *CMSpSp*

and the schema

RMS: ⊢*CSpCSq* . . . *CNMNCαβCMαMβ*,
where *p*, *q* . . . are all the variables in β that are not in α.[1]

What postulates for *Ta*, consistent with the intuitions underlying the system Q, would yield everything in Q preceded by an arbitrary *NTaN*? Note, first, that I do not say 'preceded by an arbitrary *Ta*'; this will not do because even logically true formulae of Q will contain variables which could stand for propositions which just do not occur in our arbitrary world *a*, so that the whole will not be true in *a*. However, a logically true formula will not be *false* in any world, and so should be provable in the required T-calculus preceded by *NTaN*. (In the T-calculus for S5, the prefixes *Ta* and *NTaN* are of course equivalent, by T1.) In a preliminary discussion of this problem[2] I suggested that the required T-calculus would contain both *CTaCpqCTapTaq* and *CTaKpqKTapTaq* and their converses, but in fact the converse of the former, i.e. *CCTapTaq*

[1] A proof of an established set of postulates from these ones can easily be extracted from *Past, Present and Future*, pp. 155–6.
[2] Ibid., p. 157.

TaCpq, is undesirable, for the substitution *q/p* in this would yield *CCTapTapTaCpp*, and so (by detachment of the tautological antecedent) *TaCpp*; but in fact *Cpp* would not be true (or false) in a world *a* in which *p* did not occur. *TaCpp* can be regarded precisely as stating that *p* does occur in *a*, i.e. 'It is statable in *a* that *p*'. It will be convenient in what follows to abbreviate this to *Sap*, the distinction between this *S* and the monadic *S* which figures in Q itself being gathered from the context (e.g. in '*SaSp*' the second *S* is the *S* of Q and the first *S* our defined *T*-function, the whole abridging '*TaCSpSp*').

The required T-axioms are

QT1. *ETapKSapNTaNp*
QT2. *ETaKpqKTapTaq*
QT3. *ETaMpKSapΣbTbp*
QT4. *ETaSpΠbSbp*
QT5. *ESaNpSap*
QT6. *ESaKpqKSapSaq*
QT7. *ESaMpSap*
QT8. *ESaSpSap*.

Of these, QT1 is equivalent to the three implications *CTapNTaNp* (= *CTaNpNTap*, half of T1 in the S5 T-calculus), *CTapSap* and *CSapCNTaNpTap* (= *CSapCNTapTaNp*, a qualified version of the other half of T1). QT2 is exactly the same as T2 in the S5 T-calculus. QT3 is equivalent to the three implications *CTaMpΣbTbp* (half of T3 in the S5 T-calculus), *CTaMpSap* and *CSapCΣbTbpTaMp* (a qualified version of the other half of T3—the qualification is needed because even if *p* is true in some world, it might not be true in *a*, because it might not be statable in *a*, that *p* is possible). In QT4 the conjunct *Sap* is not needed on the right-hand side, since it follows from what is already there. From this basis we can prove *CCSapTapNTaNp*, *CTaCpqCTapTaq*, *CSaqCNTaNCpqCTapTaq*, and with their help all of Q preceded by *NTaN*.

In both systems we could let the propositional variables stand for tensed propositions and the world variables for instants at which these are true. *Mp* will then of course mean, 'At some time *p*'. The truth-conditions of the ordinary tensed forms *Fp* ('It will be that *p*') and *Pp* ('It has been that *p*'), and in the Q-like case of the forms *Tp* ('It will always be statable that *p*') and *Yp* ('It has always been statable that *p*'), can then be given in obvious ways by introducing the form *Uab* for '*a* is earlier than *b*'.

In the T-calculus for S5, the symbol *T* is dispensable. We could simply regard modal propositions as *predicates* of the worlds 'in' which they are said to be true, and tensed propositions as predicates of the instants 'at' which they are said to be true. The equivalences T1–3 can then be transformed into definitions of complex predicates in terms of propositional complications, thus:

T1. $(Np)a = -(pa)$ Df.
T2. $(Kpq)a = (pa \,\&\, qa)$ Df.
T3. $(Mp)a = \exists b pb$ Df.

(Here we use Łukasiewicz symbols for functions of predicates, and symbols of a semi-Russellian sort for functions of propositions.)

An analogous interpretation of the T-calculus for Q is not so easy, since if we try it out we find that QT1 does not equate $(Np)a$ with $-(pa)$, but merely has the former implying the latter; similarly with QT3. $(Np)a$ is indeed equated with $Sap \,\&\, -(pa)$, i.e. $(Cpp)a \,\&\, -(pa)$, but here the first conjunct still has a predicative rather than a propositional complication. I have sometimes used this fact to argue that the modal or tensed propositions of Q cannot be regarded as predicates at all;[1] but there is another possibility, recently aired by Storrs McCall,[2] namely that modal or tensed propositions may be regarded as predicates in a 'non-standard' or 'non-classical' type of predicate logic. I want now to explore this possibility a little further.

McCall mentions in particular the 'free logic' of Karel Lambert. This is not quite what is needed to cope with the present problem, but I shall mention to begin with a system, due to Hughes and Londey, which resembles 'free logic' in being designed primarily to find a place for 'empty names'.[3] In this system (which at this

[1] See, e.g., *Papers on Time and Tense*, pp. 132, 142–4.

[2] S. McCall, review of *Past, Present and Future* in *Dialectica*, vol. 6, no. 4 (March 1968).

[3] See *Past, Present and Future*, pp. 168–9.

point looks promising for our present purpose) the direct application of a predicate to a non-existent object is taken to be meaningful but always false, but the system is saved from the contradiction which would arise from having both pa and $-pa$ sometimes false, by distinguishing between $(Np)a$, which is false along with pa when a is non-existent, and $-(pa)$, which under these circumstances is true, along with $-((Np)a)$. In such a system $(Cpp)a$ could be used to define 'a exists', $E!a$, and we would have such postulates as

H1. $pa \equiv (E!a \ \& \ -(Np)a)$
H2. $(Kpq)a \equiv (pa \ \& \ qa)$

corresponding to QT1 and QT2, H1 being equivalent to the three implications $pa \supset E!a$, $pa \supset -(Np)a$ and $E!a \supset (-(Np)a \supset pa)$, the latter two being equivalent to $(Np)a \supset -pa$ and $E!a \supset (-pa \supset (Np)a)$. Hughes and Londey do not, in developing this system, use the form $\exists apa$, but just write $\exists p$ to indicate that p is instantiated. They show that although we have $pa \supset \exists p$, i.e. $pa \supset \exists bpb$, we do not have $\forall p \supset pa$, where $\forall p = -\exists(Np)$; i.e. we do not have $-\exists(Np) \supset pa$, i.e. $-\exists b(Np)b \supset pa$. Whether with an expanded symbolism we would have $(-\exists b - (pb)) \supset pa$, they do not consider, but I see no reason why we should not; $\exists b - (pb)$ would mean that something, maybe something that doesn't exist, fails to be p; its negation would mean that nothing, not even any non-existent thing, fails to be p (which would have as a side-consequence that nothing fails to exist), and this does seem to entail pa for any arbitrary a. The oddities of the non-existent are not here carried by the quantifiers but by the predicates; they are partly reflected in the form $\forall p$ because the negative predicate Np is implicitly contained in this.

But while H1 and 2 above might do as interpretations of QT1 and 2, this system for empty names has something much stronger than the rewritten versions of QT5 and 6, i.e. than $(CNpNp)a \equiv (Cpp)a$ and $(CKpqKpq)a \equiv (Cpp)a \ \& \ (Cqq)a$; namely $(Cpp)a \equiv (Cqq)a$. This equivalence of *all* predicates of the form $C\alpha\alpha$ is precisely what justifies the abridgment of $(Cpp)a$ to $E!a$, in which no predicate variables appear. We do not of course have $(Cpp)a \equiv (Cqq)a$ in QT, since Cpp could be true in a but Cqq not true because not statable. What we have in QT is something more like a modification of predicate calculus to cope with a rather different problem, namely that of certain types of predicate being 'inapplicable' to certain types of subject (e.g. colour-predicates to numbers). What we are now after is a 'many-sorted' predicate logic in which the burden of the many-sortedness is not carried by the name-variables we use in quantifiers but by our method of forming complex predicates. We say that a negative predicate *Np* is 'applicable' to a subject *a* if and only if *p* is, and a conjunctive predicate *Kpq* if and only if both *p* and *q* are. We do not say, however, that the attachment of a predicate to a subject to which it is inapplicable results in something which is 'senseless' in the sense of being ill-formed, non-propositional or without a truth-value, but only that it invariably results in something which is false. Thus *pa* may be false either because *p* and *Np* are both applicable to *a* but it is (*Np*)*a* which is true, or because *p* is inapplicable to *a*, in which case *Np* will also be inapplicable, and we will have both —(*pa*) and —(*Np*)*a*. For example, we might say that the number 4 is not prime because, although primeness and non-primeness are applicable to numbers, 4 is in fact non-prime; but 4 is not blue (it is not the case that it is blue) for a different reason, namely that blueness is not applicable to numbers, and for this reason 4 is not non-blue either (it is not the case that 4 is non-blue). We can call this second sort of falsehood 'senselessness' if we like, so long as we get its logic right, and don't deny that the negation of such 'nonsense' is meaningful and even true.

The predicate *Mp* is true of a subject *a*, i.e. we have (*Mp*)*a*, if and only if *p* is applicable to *a* and is non-empty; for example 4 is 'possibly prime' because 4 is a number (and primeness is therefore applicable to it) and some numbers are prime; my tie is not possibly prime because primeness is not applicable to it, and not possibly both round and square because, although both roundness and squareness, and therefore roundness-and-squareness, are applicable to ties, nothing to which shape-predicates are applicable (nor, of course, anything else) is both round and square.

If we adopt this interpretation, it might seem possible to take *Sap*, '*p* is applicable to *a*', as an undefined function, and replace QT1–4 by the following definitions:

QT1. $(Np)a = Sap \mathbin{\&} -(pa)$ Df.
QT2. $(Kpq)a = pa \mathbin{\&} qa$ Df.
QT3. $(Mp)a = Sap \mathbin{\&} \exists bpb$ Df.
QT4. $(Sp)a = \forall bSbp$ Df.

and add $pa \supset Sap$ to the postulates for S. That $(Cpp)a$, i.e. $(NKpNp)a \equiv Sap$, would be provable from the definitions and the other postulates. These definitions would not, however, be completely eliminative, as they would not eliminate N, K, etc. from such forms as $SaNp$, $SaKpq$, etc., but only from $(Np)a$, $(Kpq)a$, etc.

A 'categorial' predicate might be defined as one which is true of every subject to which it is applicable; or in symbols, $\text{Cat}(p) = \forall a(Sap \supset pa)$. For example, 'is a number' is true of every subject to which it is applicable. From this it follows that, although plenty of things are not numbers, nothing is a non-number. My tie, for example, is not a number, but because being-a-number is not applicable to it, it is not a non-number either, and it is not a possible number or a possible non-number. Only things which could be numbers are either numbers or non-numbers, and nothing which could be a number is a non-number, or even 'not a number'; whatever could be a number is one. In symbols, we can prove $\text{Cat}(p) \supset -(MNp)a$ and $\text{Cat}(p) \supset ((Mp)a \supset pa)$. We do not, however, have $\text{Cat}(p) \supset (NMNp)a$, or even $\text{Cat}(p) \supset pa$; attributions of categorical predicates may be false, even though attributions of their negations are never true. Such attributions are like the 'structural propositions' which Johnson introduced as optional extra premisses in his expanded version of certain syllogisms.[1]

It is not difficult to present a predicate calculus of this sort as a set-theory, though this presentation raises a new problem. We can read 'pa' as 'a is in the set p', and associate each set with another set in which it is properly or improperly included, which we may call its A-set. The set Np is then not the complement of p with respect to the entire universe over which the name-variables range, but only with respect to its A-set; i.e. the set Np consists not of everything whatever that is not in p, but of all the members of p's A-set which are not in p. If the set p is empty so is the set Mp, but if the set p is non-empty the set Mp coincides with p's A-set. As far as the formal development of the theory is concerned, A-sets

[1] W. E. Johnson, *Logic*, II.i.6.

can be arbitrarily selected from sets including given sets. For example, the A-set of the set of policemen could be the set of individuals who are either policemen or door-knobs; we would then say that while many things are-not policemen, only door-knobs (i.e. things that are either-door-knobs-or-policemen but not policemen) are non-policemen. And with this assignment of A-sets, all door-knobs are possible policemen and all policemen possible door-knobs. Some sets are no doubt more 'natural' than others for use as the A-set of a given set, but this consideration lies outside the formalism. It is arguable that only an *irreducible* use of modal words will give us a sense in which, *not* just relatively to an arbitrary assignment of A-sets, a fireman is a possible policeman but a door-knob is not.

Leaving aside this last small question, the use of 'possibly' and 'could' which is illustrated by the above explanation of '4 is not but could be prime, but my tie could not be prime' is one which might be expected to commend itself to those who wish to give an extensional sense to modal expressions. Moreover, the ordinary use of modal expressions could be represented as a special case of this one. If the object x is absent from the world a, this will mean that such a predicate as 'x-eating-chocolate-ish' is inapplicable to a, so that a is neither an x-eating-chocolate-ish world nor an x-not-eating-chocolate-ish one, i.e. it is neither p-ish nor (not-p)-ish where p is 'x is eating chocolate'. A world a is an x-possibly-eating-chocolate-ish world if and only if (i) the predicate 'x-eating-chocolate-ish' is applicable to it, i.e. x exists in a, and (ii) some world (a or another) to which 'x-eating-chocolate-ish' is applicable is one which is in fact x-eating-chocolate-ish. This is how we express in predicate-calculus terms the ruling that it is the case in world a that x is possibly eating chocolate if and only if (i) it is statable in a that x is eating chocolate, and (ii) in some world in which this is statable it is true. A similar representation is possible for tenses. An instant a is an x-at-some-time-eating-chocolate-ish instant if and only if (i) the predicate 'x-eating-chocolate-ish' is applicable to the instant a, i.e. x exists at a, and (ii) some instant to which 'x-eating-chocolate-ish' is applicable is one which is in fact x-eating-chocolate-ish, i.e. one at which x is eating chocolate.

I have said that modalities and tenses *could* be represented in this way. Whether they ought to be, for any purposes but the development of formal analogies, is another question. That modal systems can be given an extensional interpretation is now a commonplace; and as we have just seen, it is true even of the system Q. But I wonder whether anybody wants to put forward anything like the following as a piece of serious metaphysics: There really are such objects as possible worlds, and what we loosely describe as propositions of modal logic are in fact predicates of which these objects are the subjects. For example, to say that grass could have been pink is to say that there is—there really is—a world in which it *is* pink. To say that grass is green, without any modal qualification, would of course be, on this view, to predicate grass-being-green-ishness of the actual world, but this word 'actual' must not be taken as signifying that the world in question is in any way more 'real' than those other worlds in which grass is pink or purple. The word 'actual' must be regarded as having the concealed egocentriticy which the temporal 'present' is sometimes said to have; the 'actual' world is just the world in which *we* figure. And even this is not quite right. For statements like, for example, 'I might have been a railwayman' must be taken to mean that there is—there really is—a world in which I *am* a railwayman. We all as it were perform on several stages at once, and in each world we are largely ignorant of our performances—real though they be—in others. I say 'largely' ignorant because our knowledge of what we might have been doing would on this view be the way in which we are obscurely aware in one world of what we are doing in others. So the 'actual' world is not strictly identifiable as 'our' world—there are other worlds that are that too (though perhaps not *all* worlds are that). The 'actual' world is just the one we indicate by waving our arms about in a vague way and calling it 'this' world. Alternatively we could follow a suggestion of David Lewis[1] and say that each of us figures in one world only but may have a 'counterpart' or counterparts in other worlds, and that our knowledge of what we might have been doing is an awareness in one world of what our counterpart is doing in another.

In either of its forms, this seems a tall story, and as I have said, I doubt whether anyone seriously believes it. But plenty of people believe an exactly similar story about tenses, i.e. believe that tensed

[1] David K. Lewis, 'Counterpart Theory and Quantified Modal Logic', *Journal of Philosophy*, vol. 65, no. 5 (March 1968).

propositions are predicates of 'instants', and that to say that, for example, I have been drinking, is to say that there is—really is—an instant at which I unalterably 'am' drinking. I do not see much more reason for believing this story than the other one; but I must confess that there are fewer knock-down arguments against it—or for that matter against the taller story about modality—than I once thought; in particular a Q-like tense-logic is as amenable to this type of interpretation as less subtle ones are. In doing metaphysics there is still no substitute for 'the choice of the soul'; or, if you like, prejudice. In any case the modalities of Q are interpretable as qualifications of predicates in the kind of predicate logic that I have described (the theory of 'applicability'). John Lemmon once said to me, not long after the system Q had been invented and he was at work on its semantics, that while this system could certainly be defended it could also be trivialised; he offered me an 'awful' interpretation of Q to be set beside what Łukasiewicz had called my 'awful' interpretation of his Ł-modal system as a logic of deliberate ambiguity.[1] I think Lemmon had in mind some such interpretation as the one I have sketched here.

All the same, a certain amount of argument, to bolster up the prejudice, is still possible here. Suppose we introduce ordinary individual variables, with quantifiers binding them, and also an identity function with such variables as arguments, into our modal or tense logic, i.e. suppose we have not merely a modalised or tensed propositional logic but a modalised or tensed predicate logic with identity, still along the lines of Q. The T-calculus for this would have to have some additions, and if we abridge *TaTxx* to *Sax*, the following suggest themselves:

QT9. *ETaΣxfxΣxTafx*
QT10. *ETaIxyKKSaxSayIxy*
QT11. *ESafxKSaxSaΣyfy*.

I shall have something to say about QT9 and 10 shortly. QT11 states that *fx* is statable in the world *a*, or at the instant *a*, if and only if this is not prevented either by some fault in *x* or by some fault in *f*. The fault in *x* would be *x*'s non-existence in or at *a*, and

[1] See my *Time and Modality*, p. 4, and article 'Logic, Many-valued' in the *Encyclopedia of Philosophy* (New York 1967), vol. 5, p. 4.

this can be represented as the failure in *a* of some formula, say *Ixx*, in which *x* is the only variable; hence we define *Sax*, *x*'s existence in *a*, as *TaIxx*, analogously to our definition of *Sap* as *TaCpp*. The fault in *f* would be its containing an individual variable with the preceding fault, but formally it would amount to the failure in *a* of some logical law in which *f* is the only free variable, e.g. *CΣyfyΣyfy*. So the absence of such a fault is represented in QT11 by *SaΣyfy*, i.e. *TaCΣyfyΣyfy*. These postulates would yield such desirable theorems as *NTaN*(*CfxΣyfy*) and the schema *NTaN* (*CKαΣxfxΣxKαfx*), where *x* is not free in α. Also, in identity theory they would yield the desirable *ETafxΣyKTaTxyTafy*. Further, these postulates would *not* yield such *un*desirable consequences as the Barcan formula *CMΣxfxΣxMfx* preceded by *NTaN*. We do indeed have, by ordinary quantification theory, *CΣaΣxTafxΣxΣaTafx*, and from this by QT9 *CΣaTaΣxfxΣxΣaTafx*, which in the T-calculus for S5 would be equivalent to *CTaMΣxfxΣxTaMfx*, and QT9 would equate this to *CTaMΣxfxTaΣxMfx*, and so to *TaCMΣxfxΣxMfx*. In the T-calculus for Q, however, some of these moves are blocked. In words,

(1) (Possibly something *f*'s) is true at *a*
= (2) It is statable at *a* that something *f*'s, and (something *f*'s) is true at some *b*
= (3) It is statable at *a* that something *f*'s, and something (*f*'s at some *b*),

where we go from (1) to (2) by QT3 and from (2) to (3) by QT9. But

(4) (Something possibly *f*'s) is true at *a*
= (5) Something (possibly-*f*'s at *a*)
= (6) Something (exists at *a*—where it is statable that something *f*'s—and *f*'s at some *b*),

and this is not deducible from (3). The point is that it may be statable at *a* that something possibly-*f*'s (and so—this does follow—there may be something of which it is statable at *a* that it possibly-*f*'s), but with regard to the thing that does possibly-*f*, i.e. *f*'s at some *b*, it may not be statable at *a* that *that* thing possibly-*f*'s, and so not be *true* of anything at *a* that it possibly *f*'s, i.e. not true at *a* that something-possibly-*f*'s—even though it is true and statable at *a* that there is a *b* where something *f*'s (the particular something in question not being mentionable at *a*).

When, however, we attempt to regard this extension of QT as part of a predicate calculus of the sort earlier sketched, we encounter serious conceptual difficulties. Let us begin, as before, by re-symbolising our postulates, thus:

QT9. $(\Sigma xfx)a \equiv \exists x(fx)a$
QT10. $(Ixy)a \equiv (Sax \;\&\; Say \;\&\; (x = y))$
QT11. $Safx \equiv (Sax \;\&\; Sa\Sigma yfy)$.

Here *f* must be interpreted as a *two*-place predicate from which we form the one-place predicate *fx* by filling the first of its places with the name *x*, and the one-place predicate Σxfx by binding a variable in its first place. *Safx* will mean that *fx* is 'applicable' to *a*, i.e. that *a* is a suitable second argument for *f*. QT9, on this interpretation, presents no problems. To say that 4 has the property of being divisible-by-something is equivalent to saying that for some *x*, 4 has the property of being divisible-by-*x*. (Here we read *fxa* as '*a* is divisible by *x*', making *f* the two-place predicate 'goes into', *fx* the one-place predicate 'is divisible by *x*', and Σxfx the one-place predicate 'is divisible by something'). But QT10 and 11 are not so easy. QT11 states that *a* is a suitable second argument for *f* provided that there is nothing about the other argument and nothing about *f* that prevents it. One can see how there could be trouble about *f*; suppose, for example, *f* were 'shaves' and *a* the number 4. Neither 'shaven', i.e. 'shaved by someone', nor 'unshaven', i.e. 'shaved by no one', seems applicable to 4, and this, as QT11 states, makes 'shaved by *x*' also inapplicable to 4. But could there be any trouble about *x*, that is to say about *x in itself*, or purely in relation to *a*? *x* might of course be an unsuitable *first* argument for *f*; but this would be expressed, if at all, by *Sxfa*, where *f* is the converse of the relation you first thought of; in fact the system has no way of expressing unsuitability of predicates relative to *x*'s, only relative to *a*'s. The conjunct *Sax* in QT11 says nothing about *x* in relation to *f* or to *fa* but only about *x* in relation to *a*; and its negation —*Sax* would seem to mean that *no* relation with *a* for one of its terms could have *x* for the other. But do we ever have cases of this? There are obvious relations between the number 4 and, for example, my bed, e.g. the relation which consists in my bed having 4 legs; and it is hard to think of any pair of terms for which *some* applicable relating predicate could not be found. But we cannot do without *Sax* when QT is being used as a T-calculus for the modal system Q; the whole motivation for Q lies in *fx* being unstatable at *a* if *x* does not exist at *a*, i.e. if —*Sax*; the component *SaΣyfy* is only necessary to exclude from statability at *a* an *fx* in which the *f* contains a *z* for which —*Saz*.

Formally, *Sax* is a definitional abbreviation for (*Ixx*)*a*, but this raises another problem when we treat QT as a predicate calculus. How can the self-identity of *x*, or the identity of *x* and *y*, be regarded as a predicate of anything at all? How can identity, in other words, be assigned an additional argument beside its obvious ones, or treated as relative to some further object? This, of course, throws doubt on the very intelligibility of QT10.

When, on the other hand, we take QT9–11 seriously as part of a theory of worlds or instants, QT11 is not only (as we have seen) perfectly in order, but is the *only* one of the three postulates which is so. I don't mean that QT9 and 10 yield any theses of modalised or tensed predicate logic which are undesirable, though it may be that they do; but they are in themselves open to intuitive objection. Let us look first, from this point of view, at QT9; and to bring out the issues more clearly, let us first suppose that the interior calculus is a tensed rather than a modal one. QT9, as re-stated, employs an interior and an exterior existential quantifier, Σ and ∃. The former is part of the tensed proposition Σ*xfx*, 'Something *f*'s', about which we can ask '*When* is this the case?' and 'Is it the case at the instant *a*?'. This second question amounts to 'Is there at the instant *a* anything which *f*'s?'. It would be unnatural to regard an affirmative answer to this question as justified when *f* is 'will fly to Sirius', if *something which does not exist at a* is going to do so. Indeed, QT9 itself rules this out; 'Σ*x*(*x* will fly to Sirius)' is by QT9 true at *a* if and only if for some *x* it is true at *a* that *x* will fly to Sirius, and this is not true at *a*, since it is not even statable at *a*, if *x* does not exist at *a*. The quantifier Σ*x*, in short, is a 'tensed' quantifier. But not ∃*x*. If we read ∃*x* this way in QT9, i.e. if in effect we replace it there by Σ*x*, that formula becomes simply false (considered as assertable of any arbitrary a), for it will imply that whenever $(\Sigma x f x)a$, i.e. whenever it is true at a (which might not be now) that something f's, then $\Sigma x(fx)a$, i.e. there is *now* something which f's at a. This reading of QT9, moreover, would make it equate a tenseless proposition with a tensed one. No, $\exists x$ belongs to the 'exterior' theory to which $\exists a$ belongs, and ranges over objects regardless of when they exist, and can be used in stating timeless relations between these objects and the times at which they exist, and indeed between them and the times at which they don't exist; for we have in the system such formulae as $\exists x(-Sax)$, asserting that there 'is' an object of which it is not statable at a that it is self-identical, or concerning which there is no such proposition at a as the proposition that it is self-identical. All the same, there 'is' the proposition that $x = x$, a special case of the form '$x = y$' which occurs as part of the 'exterior' language of QT10, which apparently can be used at any time. Or can it? There 'is' this proposition that $x = x$, but there is no such proposition at a; and a could be now.

Similarly with modality. Where the interior language is modal, $\exists x(-Sax)$ will mean that there 'is' an object which does not exist in the world a, which could be the actual world; i.e. that there 'is' an object concerning which there is no such proposition in the actual world as the proposition that it is self-identical; and yet there 'is'—but not actually!—the proposition that $x = x$, where x is this object.

These are strange ways of talking. The trouble, I think, is this: The *form* of such propositions as the right-hand side of QT9, $\exists x(fx)a$, suggests that we are not only taking 'worlds' or 'instants' seriously as objects, but taking merely possible or merely past or future ordinary individuals seriously as objects too. But the *content* of the calculus, including QT9, suggests that these x's are after all just the individuals we know, which are nothing at all except when they are actual or present, and that is not in all worlds or always, still less outside all worlds or times. So one has the feeling that something has escaped from its modal or temporal cage, and appeared where it ought not to be. Of course quantifiers and identity, like truth-functions, have their place everywhere; we must have '$\exists$' and '$=$' as well as '$-$' and '&' in the outer framework we have provided for Q, just as we must eventually introduce Σ and I as well as N, K, etc. into Q itself; but the question is, what variables should the outer '∃' bind, and the outer '=' relate? Our p's and f's belong as it were within their brackets, tied to their a's, and surely that is where our x's belong too; they have no business jumping over the fence into the timeless or super-modal framework, as they do in such forms as $\exists x(fx)a$ and the untied $x = y$. Nor do they appear there in QT11, but they do in QT9 and 10. So it seems to me that these last are a bit out of place in a T-calculus which is to preserve the ideas behind the system Q; such a calculus had better obtain its 'desirable theorems' in other (doubtless clumsier) ways.

We can indeed dissolve this distinction between the T-framework and the modal or tense logic that it encloses, and in so doing de-mythologise the former, by regarding its 'worlds' or 'instants' as modal or tensed propositions of a special sort, with special postulates, and defining Tap within the modal or tensed calculus as $MKap$ (to be true in or at a is to be possibly or at some time true along with a).[1] The distinction between '—' and N, and between '&' and K, will now disappear, and so will that between $x = y$ and Ixy. QT10 will in fact remain true on this interpretation, but will now mean that it is true *now* that Ixy is the case at a if and only if x and y both exist at a and x and y are *now* identical (a consequence of the fact that it is now statable that Ixy is the case at a if and only if Ixy is statable both at a and now). With this same interpretation, the '=' of $a = b$ will differ from the I of Ixy in being a connective rather than a two-place predicate (a and b being propositions), and $\exists a$ will differ from Σx simply in being a *higher order* quantification, just $\exists p$ with a restriction on the range of p, and it might as well be written Σa since the distinction lies not in the kind of quantification but in the kind of variable bound. But a supposedly timeless or super-modal $\exists x$ will just not be definable at all, and therefore QT9 not formulable. QT9 would in fact be true in this wholly tensed or modal system if $\exists x$ were replaced by $M\Sigma x$; but this will not do as a definition of $\exists x$, since for the latter, ordinary quantification theory gives

$$Sax \,\&\, \exists y Sby \equiv \exists y(Sax \,\&\, Sby),$$

but we do not have

[1] See *Papers on Time and Tense*, ch. XI and p. 160.

EKSaxMΣySbyMΣyKSaxSby.

For here the left-hand side means that it is now true that (*x* exists at *a* and it is at some time true that something exists at *b*), and this is compatible with nothing that exists at *a* existing at any time at which anything that exists at *b* exists. But if this last situation obtains, there is no time at which there is a *y* of which we can then truly say '*x* exists at *a* and this [i.e. *y*] exists at *b*'; which contradicts the right-hand side.

At this point, then, what is desirable in a T-calculus for Q ceases to coincide with what is desirable in a 'logic of applicability'. For the former we want QT11 and perhaps QT10 but not QT9, and for the latter QT9 but not QT10 or 11.

One further point: In *epistemic* logic, it might seem plausible to regard statements about a person's beliefs as attaching predicates to his 'belief-world'. To say that a man believes in God, for example, may be to say that God exists in his belief-world (cf. Anselm's talk of God existing 'in the mind'), and this in turn to characterise his belief-world as 'theistic'. Perhaps in another man's belief-world God does *not* exist, i.e. that belief-world is 'atheistic'. Belief-worlds are in some ways easier to swallow than possible worlds; they are reminiscent of the 'private spaces' which figure in some of Russell's theories about perception, and perhaps they can even be given a neurophysiological interpretation. But it would seem that in many cases they are characterised neither by a given predicate nor by its contradictory; in plain language, a person may neither believe that *p* nor believe that not-*p*, since he may not have thought about the matter, and even if he has may have arrived at no opinions about it. A belief-world may, for example, be neither theistic nor atheistic but agnostic. The T-calculus for Q could help us here; we could say that the predicate 'theistic', and therefore also the predicate 'atheistic', are just 'inapplicable' to certain belief-worlds, namely those of agnostics.

I am not inclined, however, to carry this line of thought very far, tempting as it is. For there are the following reasons against it:

(i) There are matters on which a person's beliefs may be not merely unformed but inconsistent, i.e. we do not have the analogue of *CTaNpNTap*, which would come from QT1. A much more serious modification of predicate logic than any we have so far contemplated would be needed to allow cases in which the predicates p and Np are not merely false of the same subject, but true of the same subject.

(ii) We often need to compare a person's belief with what actually is the case; in particular, when we say that a person's belief that p is mistaken, we mean that he believes that p but it is not the case that p, and we mean that what he believes is *precisely* what is not the case. Hence if p is genuinely propositional in 'It is not the case that p' it must be genuinely propositional, and not merely predicative, in 'x believes that p'. To say anything less is to fail to take error seriously. Hence, conversely, if p in 'x believes that p' is to be construed as merely a predicate of a belief-world, so must it be on its own, or negated; but of *which* belief-world? We cannot identify the unqualified p or Np with its being true of a selected one among a number of belief-worlds, as we might identify it with its being true in one of a number of possible worlds, or at one of a number of instants. The actual world isn't 'this' belief-world, in any sense of 'this'. At this point, so far as I can see, this theory can only be saved by taking one of two unplausible courses. We could introduce a new idealistic argument for theism, postulating a being with whose belief-world the actual world can be definitionally identified, i.e. defining the Truth as what God believes. Or we can say that there is no truth against which our beliefs are to be measured, and that simple assertion of a proposition is simply predicating it of *our* belief-world.

One might argue that there *is* one way of escaping these alternatives. There is nothing in the suggested representation of propositions about beliefs which ties the belief-worlds to *people*—they are just there, entities that we characterise. And one might introduce the actual world, call it n, as another entity of the same broad sort as the belief-worlds (n is substitutable in theorems for belief-world-variables), but with special features, e.g. though we do not in general have $-(pa) \supset (Np)a$, we do have $-(pn) \supset (Np)n$. (This brings out nicely the determinateness of what is actual by contrast with what is not.) 'a's belief that p is mistaken' would then come out as pa & $-(pn)$, or the equivalent pa & $(Np)n$. And we don't have to think of n either as God's belief-world or the speaker's; it isn't anybody's belief-world, but it is a world, and the belief-worlds are that too. This suggestion, however, still doesn't meet the first objection, or a third which we now state.

(iii) The system would make beliefs about people's beliefs inexpressible, since the form pa, i.e. 'a's belief-world is p-ish', *is* genuinely propositional and therefore not a possible characterisation of a belief-world. We have, indeed, the function $(Mp)a$, but this will *not* mean that a believes that someone believes that p, but only that a has an opinion about p and someone (a or another) does believe it. (Hintikka's system in *Knowledge and Belief*, which does at first sight look a bit like reducing assertions about beliefs to the ascription of predicates to belief-worlds, at this point precludes that interpretation, since in his system beliefs about beliefs are quite straightforwardly formulable.) To put this point another way: In the T-calculus for Q the second quantification implicit in $(MMp)a$ is a vacuous one; the formula is equivalent to $(Cpp)a \,\&\, \exists b(Mp)b$, and this to $(Cpp)a \,\&\, \exists b((Cpp)b \,\&\, \exists cpc)$, and this to $(Cpp)a \,\&\, \exists b(Cpp)b \,\&\, \exists b \exists cpc$, and this to $(Cpp)a \,\&\, \exists cpc$, i.e. to $(Mp)a$. In the T-calculus for S5 this is even more obvious; there $(MMp)a = \exists b(Mp)b = \exists b \exists cpc = \exists cpc = (Mp)a$. Intuitively this is right too, in tense and modal logic. 'It is true in a that it is true in some possible world that it is true in some possible world that p' is indeed equivalent to 'It is true in a that it is true in some possible world that p', and this (at least provided that p is statable in a) to 'It is true in some possible world that p'; and 'It is true at a that it is at some time true that p is true at some time' is equivalent to 'It is true at a that p is true at some time', and this (at least if p is statable at a) to 'p is true at some time'. But 'a believes that it is believed that it is believed that p' is not equivalent to 'a believes that it is believed that p', nor (as we have already noted) is this in turn equivalent, even if a has an opinion about p, to 'It is believed that p'.

7. *Supplement to 'Modal Logic and the Logic of Applicability'*

1. I want to ask a rather crude philosophical question, namely: is S5, considered straightforwardly as a modal system, the *true* modal system? Its original inventor, C. I. Lewis, was quite certain that it was not, his view being that the true modal system is the somewhat weaker one that he called S2. I shall maintain that Lewis was right in his negative contention, wrong in his positive. Nor do either S5 or S2 seem to me true when read as tense logics. For both of them contain as a theorem the formula

CMpMApq,

i.e. 'If it could be that *p* then it could be that *p*-or-*q*', or 'If at some time *p*, then at some time *p*-or-*q*'. And there seems to me to be *p*'s and *q*'s for which this formula is not true. I have already given counter-examples;[1] here is another of the same sort: The proposition

(A) At some time (either General de Gaulle is a babe in arms or I am President of France)

said by me, A. N. Prior, does not follow from the proposition

(B) At some time (General de Gaulle is a babe in arms),

because although the latter is true, the former is false. I say that the former is false because if there were a time at which the interior disjunction 'Either General de Gaulle is a babe in arms or I am

[1] See *Time and Modality*, p. 49, and *Past, Present and Future*, p. 153.

President of France' were true, this time would have to be either within the period of my existence or outside it; but during the period of my existence both disjuncts are false, and outside it (e.g. when de Gaulle *was* a babe in arms) there is just no such disjunction. I assume here that 'I' is used in (A) simply to indicate a particular individual, and that in a world and an instant from which an individual is simply absent, there are neither facts nor falsehoods about that individual.

2. So S5, as a modal or tense logic, will not do. As a contracted version of the uniform monadic first-order predicate calculus, however, there seems nothing wrong with it at all. In particular, there seems nothing wrong with the formula *CMpMApq* when all that it means is that if an individual *x* is *p* then that individual *x* is either *p* or *q*. What this suggests is that the parallel between being true *in* a world or *at* an instant, and being true *of* an individual, is not as close as it seemed at first to be; these relations don't even have the same formal laws. The basic difference, I think, is to be found in the region of the Law of Excluded Middle. Given any individual *x* and any predicate *p*, either it is true of *x* that it is *p*, or it is true of *x* that it is not *p*. But given any proposition *p* and a possible world *w*, we cannot say for sure that either it is true in *w* that *p* or it is true in *w* that not *p*, for in *w* there might just be no such proposition as the proposition that *p*, and so no such proposition as the proposition that not *p*. And similarly, given any tensed proposition *p* and instant *t*, we cannot say for sure that either it is true at *t* that *p* or it is true at *t* that not *p*, for at *t* there might just be no such proposition as the proposition that *p*, and so no such proposition as the proposition that not *p*.

3. If neither S5 nor S2 is the true modal system, what is? I would stick by the answer to this question which I gave in *Time and Modality*, namely that it is the system which I have called Q. This may be formalised by introducing not only the modal operator *M* but a further operator *S*, such that *Sp* may be read as 'In all possible worlds' (or 'At all instants') 'there is such a proposition as the proposition that *p*', or as I sometimes put it, 'It is always statable that *p*'. This has the following laws:

S1. $\vdash CS\alpha Sp$, where p is any of the variables in α.
S2. $\vdash CSpCSq \ldots S\alpha$, where $p, q \ldots$ are all the variables in α.

Between them S1 and S2 assert that a compound proposition is always statable if and only if all its parts are. And for *M* we have the following:

A1. $C\alpha M\alpha$
R1. If $\vdash C\alpha\beta$ then $CSpCSq \ldots CM\alpha\beta$, where all the variables in β fall within the scope of an *M* or an *S*, and $p, q \ldots$ are all the variables in β that are not in α.

These postulates no longer enable us to prove *CMpMApq* but only *CSqCMpMApq*, 'If it is always statable that *q*, then if at some time *p*, then at some time either *p* or *q*'. I found my exception to the unamended law by selecting a *q* which is *not* 'always statable'.

4. If Q rather than S5 is the true modal logic, it is natural to conclude that modal logic is not after all interpretable as a bit of predicate calculus. We should not, however, draw this conclusion too hurriedly. We can in fact give a predicate-calculus interpretation, of a kind, even to the system Q. We do it by introducing into uniform monadic first-order predicate calculus predicational complexities that are *not* definable in terms of propositional ones. As before, we use *N*, *K* and *M* (with functions defined in terms of these), and now also *S*, for forming complex predicates. But instead of the *definitions* of section 1.5 we have the following axiom-schemata relating the two sorts of complexity:

Q1. $\varphi x \rightarrow (C\varphi\varphi)x$
Q2. $(N\varphi)x \leftrightarrow (C\varphi\varphi)x \;\&\; -\varphi x$
Q3. $(K\varphi\psi)x \leftrightarrow x \;\&\; \varphi x$
Q4. $(M\varphi)x \leftrightarrow (C\varphi\varphi)x \;\&\; \exists x\varphi x$
Q5. $(S\varphi)x \leftrightarrow \forall x(C\varphi\varphi)x$
Q6. $(C\varphi\varphi)x \leftrightarrow (Cpp)x \;\&\; (Cqq)x$ etc., where p, q, etc. are all the variables in φ.

These postulates (which are equivalent to the QT1–8 of 'Modal Logic and the Logic of Applicability') clearly will not serve to define predicational complexities in terms of propositional ones, since most of them have predicational complexities on the right as well as the left. To be quite meticulous, the situation with each schema is as follows: Q 1 could not be replaced by or read as a definition, since it is not an equivalence but an implication. Q2, which expands by the definition of C to

$$(N\varphi)x \leftrightarrow (NK\varphi N\varphi)x \ \&\ -(\varphi x),$$

will not do as a definition of N in terms of '—', or in terms of anything else, since it has N on both sides. Q 3 looks as if it could serve as a definition of K in terms of '&', but in fact it does not enable us to eliminate K in favour of '&' when the former is governed by an N. For, by Q 2,

$$(NK\varphi\psi)x \leftrightarrow (NKK\varphi\psi NK\varphi\psi)x \ \&\ -(K\varphi\psi)x,$$

and here, although Q 3 turns the second conjunct into $-(\varphi x \ \&\ \psi x)$, with K eliminated, the final conjunct still has K's inside N's, and further applications of Q 2 will obviously leave us in the same position. Q 4 and Q 5 *can* be regarded as or replaced by definitions of M and S, but not in terms of quantifiers and propositional complexities alone, but in terms of these plus the undefined predicational complexities K and N. Q 6 has N and K on both sides.

4. Within the predicate calculus determined by Q 1–Q 6, an interpretation of the modal calculus Q can be given. The relation of Q to the uniform monadic first-order predicate calculus asserted by Q 1–6 is not, however, so straightforward as that of S5 to that calculus enriched by the definitions in Chapters 1–5. In the first place, it is not possible to use Q 1–6 to equate all formulae, and therefore all theorems, of the uniform monadic first-order predicate calculus to formulae of the form φx, where φ is a complex predicate. In fact there are no theorems of that calculus, enriched by Q 1–6, which are of the form φx. For example, $px \rightarrow px$ is a theorem, but it does not follow from this that $(Cpp)x$ is, and in fact $(Cpp)x$ is not. So we cannot say: φ has the form of a theorem of the modal calculus Q if and only if φx is a theorem of uniform monadic first-order predicate calculus enriched by Q 1–6; for some

φ's are theorems of φ, but no φ's meet the second condition. However, we do have the following: φ *has the form of a theorem of Q if and only if*

$$(C\varphi\varphi)x \rightarrow \varphi x$$

is a theorem of uniform first-order predicate calculus plus Q 1–6. For example, *Cpp* is a theorem of Q, and

$$(CCppCpp)x \rightarrow (Cpp)x$$

follows immediately from Q 6. Alternatively we may say, as in Chapter 6, that φ has the form of a theorem of Q if and only if —$(N\varphi)x$ is a theorem of first-order predicate calculus enriched by Q 1–6, i.e. if it is provable of any object that it is not non-φ. (This condition is equivalent to the other.)

5. So far this is a purely formal result. But if the predicational complexities which figure in Q 1–6 are not definable in terms of propositional ones, what do or could we mean by them? It is, indeed, not at all difficult to interpret these postulates in terms of truth in worlds or at instants. Anything of the form *C*φφ is true in a world or at an instant if and only if φ is statable in that world or at that instant. So Q 1 states that if φ is true in or at *x* then it is statable there; Q 2 that not –φ is true in or at *x*, if and only if φ is statable but not true there; Q 3 that φ–and–ψ is true in or at *x*, if and only if both φ and ψ are; Q 4 that it is true in or at *x* that possibly φ, if and only if φ is statable there and true somewhere; Q 5 that it is true in or at *x* that φ is always statable, if and only if it *is* always statable; and Q 6 that φ is statable at *x* if and only if all its components and propositions are. And there are no theorems of the form φ*x* because however tautological φ might be, it might fail to be true in or at *x* because it might be unstatable there. The most we can have is that it is true wherever it is statable, i.e. $(C\varphi\varphi)x \rightarrow \varphi x$. But what interpretation can we give to Q 1–6 in terms of predicates being true of individuals?

6. It would not be entirely a new thing to suggest that predicational complexity cannot be wholly defined in terms of propositional. Just that suggestion could emerge, for example, from an attempt to

deal with non-existent objects in quite a different spirit from my own in Section 4 above, namely by saying that even where x does not exist there is such a proposition as the proposition that px, only all such propositions are false; that is, what is non-existent has no properties. But non-p is as much a property as p, i.e. $(Np)x$ as well as px is of the form φx, so if $(Np)x$ were simply equivalent to $-(px)$, we would have to say that where x is non-existent both px and $-(px)$ are false, and this would rather complicate our propositional logic. We may avoid this awkwardness by ***not*** defining $(Np)x$ as $-px$, but instead relating the two by laws like our Q 2. A predicate of the form $C\varphi\varphi$, we might now say, is true of anything that has any predicates true of it at all, i.e. of anything that exists; and of what doesn't exist it is false; i.e. where x doesn't exist we have, not indeed $(NC\varphi\varphi)x$, since $NC\varphi\varphi$ is a predicate too, but $-(C\varphi\varphi)x$. Q 1 now says that if x is φ it exists; Q 2 that x is not $-\varphi$ if and only if it exists and it is not the case that it is ψ; and Q 3 that x is φ–and–ψ if and only if it is φ and it is ψ (here nothing need be said about its existing, since that follows from its being φ and from its being ψ, by Q 1). Q 4 and Q 5 may be regarded as defining two rather odd concepts for which we have no ordinary equivalents. By Q 4 we may say that x is $M\varphi$, or that it is *true of x that something is* φ, if and only if x exists and something is φ. And by Q 5 we may say that x is $S\varphi$, or that it is *true of x that everything exists*, if and only if everything exists. As to Q 6, the most important thing to say about it is that with the interpretation of N, K, etc. which we are now exploring, we can lay down not merely Q 6 but something much stronger, namely

H6. $(C\varphi\varphi)x \leftrightarrow (C\psi\psi)x$,

asserting in effect that if x has any property of the form $C\varphi\varphi$ it has all properties of this form (not only, as in Q 6, those $C\psi\psi$'s in which ψ is a component of φ, but equally those $C\psi\psi$'s in which ψ has no connexion with φ at all). This is true with the suggested interpretation since if x has the property C it exists, and if it exists it also has any other 'tautological' property. Indeed, we could modify Q 1–5 by replacing $C\varphi\varphi$ throughout by the constant $E!$ (reading $E!x$ as 'x exists'), thus:

Q 1′. $Qx \rightarrow E!x$
Q 2′. $(N\varphi)x \leftrightarrow E!x \;\&\; -\phi x$
Q 3′. $(K\varphi\psi)x \leftrightarrow \phi x \;\&\; \psi x$
Q 4′. $(M\varphi)x \leftrightarrow E!x \;\&\; \exists x\phi x$
Q 5′. $(S\varphi)x \leftrightarrow \forall xE!x$

and then prove H6 as follows:

$(C\varphi\varphi)x = (NK\varphi N\varphi)x$, by Df. C,
$\leftrightarrow E!x \;\&\; -(K\varphi N\varphi)$, by Q 2′,
$\leftrightarrow E!x \;\&\; -(\varphi x \;\&\; (N\varphi)x)$, by Q 3′,
$\leftrightarrow E!x \;\&\; -(\varphi x E!x \;\&\; -\varphi x)$, by Q 2′,
$\leftrightarrow E!x$, by propositional calculus,
$\leftrightarrow (C\psi\psi)x$, by a similar proof reversed.

It may be noted that, if we thus allow ourselves the predicate $E!$ and modify Q 1–5 to Q 1′–5′, the equivalences among these latter *can* be regarded as, or replaced by, definitions—definitions of the predicational complexities, not in terms of the propositional complexities alone, but in terms of these and $E!$

If, then, we modify Q 1–6 either by strengthening Q 6 to H6 or by replacing the function $C\phi\phi$ in Q 1–5 by the constant $E!$, we obtain a version of predicate calculus of which the interpretation is quite straightforward. Those strengthenings, however, entirely destroy the plausibility of this calculus as a logic of truth in possible worlds and instants. This is particularly obvious if we use the constant $E!$ Interpreted in the theory of truth in possible worlds, Q 2′ would state that not p is true in a world x if and only if there is such a world as x and p is not true in it. But the possibility of the world x not existing is just not what we are worried about. We are worried about non-existence, certainly, but not about the non-existence *of* the worlds in which our p's and q's are or are not the case; our worry is, rather, about the non-existence *in* those worlds of the objects that our p's and q's are about.

7. There is quite a lot to disentangle here. Let us call the version of predicate calculus developed in section 4 the system H. This is in a sense a rival to the system Q for dealing with the non-existent. If we hold that in a possible world from which a certain subject is

absent there are just no predications directly about that subject, our quantified modal logic might consist of a fairly orthodox predicate calculus embedded in the modal system Q. If, on the other hand, we hold that in a possible world from which a certain subject is absent all predications directly about that subject are false, our quantified modal logic might consist of a fairly orthodox modal logic (say S5) with the predicate calculus H embedded in it. As between these alternatives, I am, as I indicated in section 3, in favour of the former. But it was not in the context of this rivalry that the system H was introduced in section 6. There, it was a question, not of combining modality with quantification over ordinary individuals, but of drawing an analogy between them—of treating modality as if it were itself a kind of quantification, a quantification over worlds or times. Or to put it another way, our aim there was not to modalise predicate logic as well as propositional logic, but to *compare* modalised propositional logic with an unmodalised predicate logic, or to look at modalised propositional logic as if it were itself an unmodalised predicate logic with the predicates attaching to worlds or times. And our problem was, what version of unmodalised predicate logic would, when its predicates were attached to worlds or times, give us the modalised propositional logic Q? The answer seemed at first to be: Something like the system H. However, it became obvious that it couldn't be *exactly* the system H; and that is the point to which we have now come.

8. To allow for non-existent worlds, when our logic of non-existence is H, is to allow for worlds in which *no* proposition is true or false; i.e. of which it is always false to say that such-and-such a proposition, or its negation, is true in them. For in H the attachment of a predicate (simple or complex) to a non-existent subject is always false. But what we are after is not this, but a logic which allows for worlds of which the following is true: *There are some propositions of which it is false to say that either they or their negations are true in these worlds and other propositions of which it is true to say this.* The predicate logic that would be really analogous to this (or which would give us this when applied to worlds as subjects) would be one allowing for individuals of which the following is true: *There are some predicates of which it is false to say that either*

they or their negations are true of those individuals, and other predicates of which it is true to say this. Or (since if neither φ or *N*φ is true of an individual then *C*φφ will be), there are *x*'s for which we have (*C*φφ)*x* with some φ's and —(*C*φφ)*x* with others. This is exactly what is excluded when Q 6 is strengthened to H6. But it is now fairly obvious what sort of predicate logic is required. As was observed in Chapter 6, what we want is a predicate logic in which predicates are divided into those which are 'applicable' to a given individual and those which are not, a predicate being 'applicable' to an individual if either it or its negation is true of that individual, and 'not applicable' if neither it nor its negation is true of that individual. This is the real parallel to the division of propositions into those which are statable in a given world and those which are not, a proposition being statable in a given world if either it or its negation is true in that world, and not statable there if neither it nor its negation is true there.

9. The predicate calculus defined by Q 1–6 is not in danger of leading a double life in relation to the system Q, as the system H was; for the predicate calculus defined by Q 1–Q 2 is *only* a parallel or analogy to Q's theory of truth in worlds and times, and is not also (as H is) a rival way of handling the problem of the non-existents. Considered straightforwardly as a predicate calculus, it is just not designed to handle the problem of the non-existent at all, but to handle what might be called the problem of the inappropriate in the sense in which it might be called 'inappropriate' to say either that the number 4 is blue or that it is non-blue. Interpreting Q 1–6 this way, we may read '—(*Cpp*)*x*', not as '*x* does not exist', but as '*p* is inappropriate to *x*', or as I put it earlier, 'inapplicable' to *x*. If we are to abbreviate the form (*Cpp*)*x*, it cannot be to anything like *E*!*x* from which the reference to the specific predicate *p* (an indifference which only H6 would justify), but rather to something like *Sxp* ('*p* is suitable for *x*', one might say), in which the *p* still occurs. The meanings of Q 1–6, considered still as defining a predicate calculus, would then be as follows: Q 1: If φ is true of *x* it is suitable for *x*. Q 2: Not-φ is true of *x* if and only if φ is suitable for *x* but not true of it. Q 3: φ-and-ψ is true of *x* if and only if each of φ and ψ is. Q 4: *x* is possibly –φ if and only if φ is suitable for *x* and true of something. Q 5: *x* is

universally—suitably—φ if and only if φ is suitable for everything. Q6: A complex φ is suitable for *x* if and only if all of its components are. As noted in section 4, Q2 and Q3 cannot be regarded as, or replaced by, definitions of *N* and *K* in terms of — and &, but Q4 and Q5 can be regarded as, or replaced by, definitions of *M* and *S* in terms of *N*, *K* and ∃. And Q4, considered as a definition of the predicate 'possibly-φ', is from some points of view quite attractive. We sometimes *do* say that a thing is possibly-φ, or could be φ, or might have been φ, when we mean that it is the sort of thing which it is appropriate to describe as being φ or being not-φ, and some things of that sort are φ. Q5 is interesting too; it amounts to asserting that φ is the sort of predicate that has no limits to its applicability—perhaps, in the medieval sense, a 'transcendental' predicate. 'I am thinking of —' and 'I like —' might be offered as examples; it is, it might be argued, as appropriate to say 'I am thinking of the number 4' as it is to say 'I am thinking of my dog'.

10. So it would be formally possible to regard the system Q as collecting together those predicates which can be proved *true of whatever individuals they are applicable to*, in a predicate calculus supplemented by Q1–6. Or: Q as a modal logic or tense logic can be regarded as the application of such a 'logic of applicability' to the predicates of possible worlds or of instants. I am not myself very happy about this; some of the reasons I have given in Chapter 6, and to those I would now add a very simple one that the 'logic of applicability' itself, i.e. predicate calculus plus Q1–6, doesn't seem to me a particularly good instrument for its purpose (just as the system H of section 9 seems to me a poor instrument for *its* purpose, the handling of the non-existent). Certainly 'Virtue is tall' suffers from a different defect from 'Dr. Dolfuss was tall', but I think we can tell a better story about this than the one that 'Virtue is tall' is false because height predicates are inapplicable to moral qualities, and 'Dr. Dolfuss was tall' is false because the height predicate that was true of him was not 'tall' but 'short'. This better story is that whereas 'Dr. Dolfuss was tall' is false, 'Virtue is tall' is ill-formed, because the only adjectives that make sentences when attached to 'Virtue' by 'is' are ones that can be paraphrased away, as when we paraphrase 'Virtue is prevalent' as 'Many people are

virtuous'. In short, although the modal or tense-logical system Q can be regarded as a fragment of a predicate logic, it could only be a fragment of a *bad* predicate logic.

11. Yet, as we have seen in section 8, the postulates Q 1–6 do have a good deal of plausibility when regarded as being about worlds or instants. This plausibility, however, seems to me readily explicable if we interpret the theory of worlds or instants, not as an applied predicate logic, but as an extended modal logic. To be more precise: Suppose we equate each instant (or world) with some always statable proposition which is true at that instant (in that world), and there only, and suppose we equate being 'true and an instant' (or 'in a world') with being somewhere true in conjunction with that instant-proposition (or world-proposition); Q 1–6 will then be deducible from the system Q plus suitable postulates for these rather special propositions. To show this in detail, we may begin by subjecting the postulates Q 1–6 to the following series of symbolic transformations:

(i) Except where otherwise stated, we replace axiom-schemata by axioms, with a rule of substitution for the variables *p*, *q*, *r*, etc. This will turn Q 1 into '$px \rightarrow (Cpp)x$' and so on.

(ii) We replace *x* by *a*, as a step towards removing the impression that we are dealing with a name-variable here. This turns Q 1 into '$pa \rightarrow (Cpp)a$'.

(iii) We replace the old forms φa by forms $Ta\varphi$. ('It is true at *a* that φ'.) This turns Q 1 into $Tap \rightarrow TaCpp$.

(iv) We drop arrows, ampersands, etc., and use Łukasiewicz-style symbolism throughout, with Π and Σ for quantifiers. This turns Q 1 into *CTapTaCpp*, and Q 4 into $ETaMpKTaCpp\Sigma aTap$.

(v) We replace the equivalence Q 2–6 by pairs or groups of implications, so that we now have

Q 1. *CTapTaCpp*
Q 2.1. *CTaNpTaCpp*
Q 2.2. *CTaNpNTap*

Q 2.3. *CTaCppCNTapTaNp*
Q 3.1. *CTaKpqTap*
Q 3.2. *CTaKpqTaq*
Q 3.3. *CTapCTaqTaKpq*
Q 4.1. *CTaMpTaCpp*
Q 4.2. *CTaMpΣaTap*
Q 4.3. *CTaCppCΣaTapTaMp*
Q 5.1. *CTaSpΠaTaCpp*
Q 5.2. *CΠaTaCppTaSp*
Q 6.1. *CTaCααTapCpp*, where *p* is any variable in α.
Q 6.2. *CTaCppCtaCqq . . . TaCαα*, where *p*, *q* . . . are all the variables in α.

Q 6.1 and Q 6.2 unlike the others, retain their schematic form, but could be replaced, as they are in Chapter 6, by a collection of axioms concerned with particular α's (e.g. *CTaCNpNpTaCpp* and its converse). The claim I am now making is the following one: Suppose we regard *a* as a special sort of propositional variable, substitutable for *p*, *q* etc. in theses (though not *vice-versa*), and the form *Ta*φ as an abbreviation of *MKa*φ; then the above postulates Q 1 to Q 6.2 are provable from the system Q plus a small group of special postulates for the special variables.

12. In detail, Q 2.1 and Q 4.1 can be dropped as following immediately from Q 1 and Q 6.1. Of the remainder, all but Q 2.2, Q 3.3, Q 4.2, Q 4.3, Q 5.1 and Q 5.2 are sample substitutions in theses of Q, i.e. Q has these formulae in the more general form in which *a* is replaced by *r*, say, and the substitution of *a* for *r* yields our postulate. For example, Q contains *CMKrpMKrCpp*, from which we have *CMKapMKaCpp*, and so by the definition of *T*, *CTapTaCpp*, i.e. Q 1. Again, Q has *CMKrCppCNMKrpMKrNp*, which when *r* is replaced by *a* and *MK* by *T* becomes Q 2.3. For Q 4.3 we need only Q and the ordinary rules of quantification, thus:

1. *CMpCMKrCppMKrMp* (in Q)
2. *CMKapCMKrCppMKrMp* (1, *CMKapMp*)
3. *CΣaMKapCMKrCppMKrMp* (2, quantification theory)
4. *CMKrCppCΣaMKapMKrMp* (3, *CCpCqrCqCpr*),

and when r is here replaced by a and MK by T we have Q4.3. For the remainder we need the following special postulates for the special type of proposition symbolised by a:

P1. $CMKaNpNMKap$
P2. $CMKapCMKaqMKaKpq$
P3. Ma
P4. Sa
P5. Σaa
P6. $C\Pi aMKaCppSp$

and the 'Barcan schema' for propositional quantification, $CM\Sigma a\alpha\Sigma aM\alpha$. I conjectured[1] that of these P1 and P3–5 would be sufficient, but while I have not since disproved this, I have been unable to obtain P2 or P6 from the others, and suspect that they cannot be so obtained.[2] Q2.2 and Q3.3 are of course just P1 and P2 with MK contracted to T, and the proofs of Q2.2, Q3.3, Q4.1, Q5.1 and Q5.2 and not very difficult, but I shall prove Q4.1 to illustrate how proofs go on in Q.

1. $C\Sigma aaCpK\Sigma aap$ ($CqCpKqp$)
2. $CNCpK\Sigma aapN\Sigma aa$ (1, $CCpqCNqNp$)
3. $CNCpK\Sigma aapMN\Sigma aa$ (2, A1)
4. $CNMN\Sigma aaCpK\Sigma aap$ (3, $CCNpqCNqp$)
5. $CCMaNa\Sigma aa$ (P5, $CpCqp$)
6. $CN\Sigma aaNCMaMa$ (5, $CCpqCNqNp$)
7. $CSaCMN\Sigma aaNCMaMa$ (6, R1)
8. $CMN\Sigma aaNCMaMa$ (7, P4)
9. $CCMaMaNMNaa$ (8, $CCpNqCqNp$)
10. $NMN\Sigma aa$ (9, Cpp)
11. $CpK\Sigma aap$ (4, 10)
12. $CpMK\Sigma aap$ (11, A1)
13. $CMpMK\Sigma aap$ (12, R1, P4)
14. $CK\Sigma aap\Sigma aKap$ (quantification theory)
15. $CMK\Sigma aapM\Sigma aKap$ (14, A1, R1)
16. $CM\Sigma aKap\Sigma aMKap$ (Barcan)
17. $CMp\Sigma aMKap$ (13, 15, 16)

[1] *Papers on Time and Tense*, p. 160.
[2] [Prior is right: they cannot be so obtained (Ed.).]

18. *CMKaMpMMp* (*CKpqq*, A1, R1)
19. *CMMpMp* (*CMpMp*, R1)
20. *CMKaMpΣaMKap* (18, 19, 17)
21. *CTaMpΣaTap* (20, D1, I).

Here 17 is a rather important lemma, and can replace P5 as an axiom. This has been shown by J. T. Canty as follows: We first prove *NMNΣaa* by *reductio ad absurdum*,

C(1)*MNΣaa*
KΣaMK(2)*a* } ((1), 17)
K(3)*NΣaa*
(4)*Σaa* (2, quantification theory).

And we have *CNMNpp* by *CNpMNp* (instances of A1) and *CCNpqCNqp*.

13. Strictly speaking, Canty's was a slightly more complicated proof of a slightly more complicated metatheorem. It is possible to dispense with special variables for worlds or instants if one introduces quantification over ordinary propositional variables into modal or tense logic, and then defines a function *Q p* meaning '*p* is an instant-proposition' (or '*p* is a world-proposition'). We can then show that the postulate *ΣpKpQ p* ('There is a true world-proposition') is deducible from the postulate *CMqΣpKQ pMKpq* ('If *p* is possible there is a possible world in which it is true'); this is Canty's theorem, proved by him in the context of propositionally quantified S5 rather than propositionally quantified Q. Propositionally quantified S5 has been quite extensively studied recently by D. Kaplan, K. Fine and R. A. Bull;[1] propositionally quantified Q, with the function *Q p* defined within it, so that special variables need not be used for propositions satisfying this function, would be the final stage in removing from Q 1–6 the appearance of belonging to predicate calculus, and planting the theory of worlds or instants firmly within modal or tense logic.

[1] R. A. Bull, 'On Modal Logics with Propositional Quantifiers', *Journal of Symbolic Logic*, 34 (1969), and R. A. Bull, 'On Possible Worlds in Propositional Calculus' (forthcoming).

8. Postscript by Kit Fine: Prior on the Construction of Possible Worlds and Instants

Fundamental to Prior's conception of modality were the following two theses:

The ordinary modal idioms (necessarily, possibly) are primitive;[1]
Only actual objects exist.[2]

The first thesis might be called Modalism or Priority, in view of its nature and founder. The second thesis is sometimes called Actualism, and the two theses together I call Modal Actualism. Prior held corresponding theses about time:

The tenses (it will be, it was the case) are primitive;[3]
Only present objects exist.[4]

His acceptance of the last thesis was tentative. For he was unsure about the existence of past individuals[5] and he was, in any case, attracted towards a more basic ontology of 'stuff'.[6]

In contrast to these doctrines about time and modality, one might hold that ordinary modal discourse is to be explained in terms of possible worlds and possible individuals and that tense-logical dis-

[1] Many references might be given. See, e.g., 'Modal Logic and the Logic of Applicability', *Theoria*, 34 (1968), reprinted as Chapter 6 above.

[2] See *Papers on Time and Tense*, p. 143.

[3] Ibid., especially ch. XII.

[4] Ibid., p. 143.

[5] See the discussion in *Past, Present and Future*, pp. 171–2.

[6] Ibid., p. 174, and *Papers on Time and Tense*, p. 80.

course is to be explained in terms of instants and individuals, past, present or future. Thus 'Possibly someone is over ten feet tall' would be analysed as 'Some possible individual is over ten feet tall in some possible world', and 'It will be that someone flies to Venus' would receive an analysis of the form 'Some person (past, present or future) flies to Venus at some time later than *t*'. These analyses are often associated with the 'possible worlds' semantics for modal and tense-logic, although there is no need to regard the semantics as providing an *analysis* of the notions that appear in the object-language.

The possible worlds analysis is incompatible with both modalism and actualism: for it analyses the ordinary modal idioms (indeed, it uses no intensional connectives at all) and it admits possibles. One could, in theory, deny either tenet without denying the other. The modalist could admit possibles and the actualist could attempt to analyse the modal idioms. However, neither of these middle positions is particularly plausible. For if the modalist admits possibles then why does he not accept the possible worlds analysis, and if the actualist does not accept the modal idioms then how, exactly, are they to be analysed?

The most plausible positions are the extreme ones and, indeed, they stand in natural opposition to one another. Both find a distinctive role for modality, but they locate it in a different aspect of language. For the modal actualist, possible objects do not exist; rather, the possible exists as a *manner* in which things happen. It exists as a mode, not an object. In the proper language for expressing modal truths, the modal primitives will be adverbial (sentential connectives) and the quantifiers will range over actual objects alone. For the classical possibilist, there is no special manner in which things happen; the possible is located in ontology. It exists as an object not mode. In the proper language for expressing modal truths, the quantifiers range over possible objects and the connectives are all truth-functional.

This ontological difference can be brought out by considering general possibilities. Take as an example, 'Possibly some individual is not actual'. For the possibilist, this is an existential claim to the effect that some possible individual is not actual. Therefore there must be some specific individual who is not actual. But for the actualist, this singularity is spurious; there can be no instance in virtue of which the sentence is true. The sentence states an irreducible general possibility, and no matter how well the individual is described, he can have no specific identity. The same considerations apply, *mutatis mutandis*, to time.

Which of these positions is correct is a large and difficult question. I am inclined to think that Prior's views on both time and modality are correct, but that the arguments for the two cases are rather different. My aim is not to discuss whether modal actualisation, or its tense-logical counterpart, is correct. Rather, it is to discuss a problem that arises once modal actualism is accepted. The problem is this: the possibilist can analyse ordinary modal discourse in terms of possible worlds and possible individuals; but what is the modal actualist to make of the possibilist's discourse?

One answer to this problem is that all such discourse is illegitimate. But this is hard to accept, for a great deal that we want to say about time and modality is put in these terms. Some of this talk may not make sense. The domain of possible worlds involves the notion of totality twice over, for each possible world is a totality (of what is the case) and then the domain is the totality of such totalities. Now it may be that certain references to these or related totalities lead to the type of paradox that is familiar in set theory. But all the same, a great deal of possible worlds (or instants) talk will still be in order.

A more acceptable answer is that this talk is legitimate, but not basic; it stands in need of analysis. The modal actualist will eliminate talk of possible worlds and possible objects in favour of the ordinary modal idioms and quantification over actuals. His tense-logical counterpart will do likewise. Thus the previous analysis will be turned on its head, with the connectives and restricted quantifiers—be they modal or tense-logical—coming first and the worlds or instants and unrestricted quantifiers emerging as a construct from them.

My aim in this paper is to carry out this programme of reconstruction, at least in outline. I have often followed the lead of Prior, much of whose later work[1] arose from this programme. However, I cannot be sure that he would have approved of all of the steps I take.

[1] See the chapters of this book, ch. XI of *Papers on Time and Tense*, and ch. V of *Past, Present and Future*.

1. *The simplest case*

The simplest case of reconstruction concerns the language of S5, i.e. the language of truth-functions and necessity. Let us begin with the possible worlds analysis of this language. The main idea behind the analysis is that the relation 'A is true in possible world w' can be expressed as a classical (indeed, first-order) condition $A(w)$. Thus 'P is true at w', for P a sentence-letter is expressed as '$P'w$', where P' is a possible worlds predicate corresponding to P; the truth-functional connectives are represented by themselves; and '$\Box B'$ is true at w' is expressed as '$B'(v)$ for all possible worlds v'. This last clause states that $\Box B$ is true at a world if B is true at every possible world. Given the condition $A'(w)$, the modal formula A can then be given the translation '$A'(\omega)$', where ω is a constant designating the actual world. Thus a formula is said to be true if it is true in the actual world.[1]

The language into which we translate is that for the monadic predicate calculus. It contains a monadic predicate P' for each sentence-letter P, a quantifier $\forall w$ over possible worlds, and a constant ω for the actual world. In order to translate back into a modal language, Prior suggested that each possible world be treated as a world-proposition. p is a *world-proposition*—in symbols $Q\,p$—if it is true in one world alone or, to put it in modal terms, if it is possible that p is true and necessarily implies all truths, i.e. $\Diamond(p \wedge \forall q(q \supset \Box(p \supset q)))$. Prior's suggestion, then, was to replace $\forall w$ with a quantifier $\forall q$ over all world-propositions, i.e. to translate $\forall wA$ as $\forall p(Q\,p \supset A^*)$ where A^* translates A. The predication $P'w$ would then be replaced by the strict implication $\Box(q \supset P)$, where q is the world-proposition corresponding to w, and the constant ω by the description 'the true world-proposition'.

The whole procedure involves three languages and two translations. There is the original or *primary* modal language for S5, the *classical* language of possible worlds, and the *secondary* modal language for S5 with propositional quantifiers. There is the *standard* translation (A to A') from the first modal language into the classical language and the *reverse* translation (A to A^*) from the classical language into the second modal language; putting these two translations together results in a third translation (A to A'^*) from the one modal language into the other. For example, suppose one

[1] See Chapter 2 above. Note that Prior does not use ω.

starts with $\Diamond P$ in the original modal language. The standard translation takes this formula into $\exists w P'w$ (= 'P is true in some possible world') and then the reverse translation takes that formula into $\exists p(Q\,p \wedge \Box(p \supset P))$, i.e. 'P is strictly implied by some world-proposition'.[1] Thus the possible worlds rendering of $\Diamond P$ is seen to rest on the fact that each possibility can be embedded in a maximal possibility.

Two aspects of the reverse translation are worth noting. The first is that the possible worlds predicate $A'(w)$ (= 'A is true in w') is translated into a rigid predicate of propositions, i.e. one that necessarily holds or necessarily fails of any given proposition. (I have used the term 'rigid' in analogy to Kripke's use of 'rigid designation'; a rigid predicate has the same extension in each possible world.) For example, each atomic formula $P'w$ is translated into a strict implication. In the reverse translation, the only source of contingency is in the reference to the world proposition. Thus $P'w$ becomes 'the true world-proposition enjoys the (necessary) property of strictly implying P'. The reference is contingent and the predication necessary, as in the well-known sentence 'the number of planets is greater than 7'. This aspect of the translation will later be exploited in eliminating quantification over possible indivduals.

The second aspect is that the correctness of the reverse translations depends upon two assumptions about the existence of propositions. The first is that world-propositions (generally, propositions) necessarily exist or, to be more accurate, that *necessarily* world-propositions necessarily exist. This assumption will be questioned later. The second is that for each possible world there is a proposition true in that possible world alone or, to put it in modal terms, that necessarily there is a true world-proposition.

Is this assumption justified?[2] Probably not, if one makes the nominalist requirement that each proposition be expressible by a single sentence. But it is not clear that such a nominalist should accept possible worlds talk in the first place. For if the existence of propositions is tied to sentences, the existence of possible worlds should be tied to their names and then there is a similar difficulty.

If one is more platonic about the existence of propositions, then justification is possible. For let X be the set of true propositions,

[1] Chapter 6 above, p. 87f.

[2] See Chapter 2 above, p. 33, for a brief discussion of this question.

and let q be the proposition that all the members of X are true. Then q is a true world-proposition. For q is true since all members of X are true, and q strictly implies all truths since all truths belong to X. To establish that *necessarily* there is a true world-proposition, note that the premisses of the above argument (that X and q exist) are necessary. Therefore the conclusion (that there is a true world-proposition) is also necessary.

Simple as this argument is, it resorts to quantification over sets of propositions and yet has a conclusion whose formulation only requires quantification over single propositions. I suspect that a language with quantifiers over sets of propositions is the natural setting for a great deal of significant modal discourse.

2. *Introducing predicates*

Let us now consider the complications that arise from adding predicates and quantifiers to the previous language, thereby yielding the language of first-order logic and necessity. I adopt the type of semantics I have adopted before[1] for this and most subsequent quantificational languages. I shall not justify this choice, but I will consider, in §7 below, the consequences of adopting the Q-type semantics of Prior. For convenience, we suppose that the original modal language contains a predicate E for existence; and to avoid notational confusion, we use $x, y, \ldots$ for the actualist variables of a modal language and $\mathbf{x}, \mathbf{y}, \ldots$ for the possibilist variables of the classical language. As before, the relation 'A is true in the possible world w' is expressed as a first-order condition $A'(w)$. Thus 'Rxy true at w', R a predicate, is expressed as $R'\mathbf{xy}w$, where R' is a possible worlds predicate (in one of its argument) corresponding to R; the truth-functional connectives and necessity operator are dealt with as before; and '$\forall xB$ is true at w' is expressed as '$B'(w)$ for every possible individual such that $E'\mathbf{x}w$'. If we let the domain of a world be the set of individuals that exist in the world, then the last clause says that $\forall xB$ is true at a world if B is true of every possible individual in the domain of the world. Again, A is given the translation $A'(\omega)$.

The language into which we translate is still first-order, but two-sorted and polyadic. It contains variables $\mathbf{x}, \mathbf{y}, \ldots$ for possible

[1] K. Fine, 'Propositional Quantifiers in Modal Logic', *Theoria*, 36 (1970), pt. 3, pp. 336–46.

individuals as well as variables $w, v, \ldots$ for possible worlds, and for each predicate in the modal language, there is a predicate with an extra argument for possible worlds in the classical language. Unfortunately, there is no automatic extension of the previous reverse translation to the new classical language. The difficulty is that possible worlds should satisfy the condition that if $\exists xB(x)$ is true in a world then an instance of $B(x)$ is true in that world.[1] So if world-propositions are to do duty for possible worlds, they should satisfy the condition that if $\exists xB(x)$ is strictly implied by a world-proposition q then an instance of $B(x)$ is strictly implied by q. But given an actualist interpretation of the quantifiers, there is no reason to suppose that this is true. Take any counter-example to the Barcan Formula, say 'Possibly there is a divine being but there is no possibly divine being'. Then there is a world-proposition that strictly implies the existence of a divine being, but there is no (actual) individual whose divinity the world-proposition implies.

One solution to this difficulty is to replace each possible individual with a property that is possibly instantiated; the property could then exist even if the individual did not. To be more specific, say that a property φ is an *individual essence*[2] *of* the (actual) individual x if φ is true of x and of x alone in each possible world or, to put it in modal terms, if necessarily for all y (φ holds of y iff y is x); and call φ an *individual essence* if it is, or possibly is, the individual essence of some individual. Then possible individuals can be replaced with their individual essences. In the reverse translation, the individual quantifier $\forall \mathrm{x}$ will be replaced with the quantifier $\forall\varphi$ over individual essences, and the predication $R'\mathrm{xy}w$ will be replaced with the formula 'Possibly there is an individual x and possibly there is an individual y such that p strictly implies φ of x, ψ of y and Rxy' where p is the world-proposition corresponding to w, and φ and ψ are the individual essences corresponding to x and y respectively. The above difficulty is now avoided, for if p strictly implies $\exists xB(x)$ then there is an individual essence φ such that possibly for some x, p strictly implies φ of x and $B(x)$.

This solution is natural. It is also consonant with the treatment of possible worlds; for in both cases, an object is identified with the

[1] Prior mentions this difficulty in Chapter 6 above, p. 95.

[2] Cf. R. Chisholm, 'Identity through Possible Worlds: Some Questions', *Nous*, 1 (1967), pp. 1–8.

sum of its necessary properties. The use of this sum again requires certain existence assumptions. The first is that individual essences (generally, properties) necessarily exist. This assumption, like the corresponding one for propositions, will be criticised later. The second assumption is that necessarily each individual has an individual essence. It can be defended in the following way. Let x be any (actual) individual and let φ be the property of being identical to x (I use identity without existential presupposition, so that each individual is necessarily self-identical). Then necessarily for all y (φ holds of y if y is identical to x); and so φ is an individual essence of x. Now necessarily for each x there is the corresponding property φ, and so the conclusion holds necessarily. This argument is, in a sense, purer than the corresponding one for world-propositions, for it requires neither a platonic stance on properties nor a resort to higher order entities.

3. *Being extensional*

In the previous reverse translations, there is apparently a gain in ontological simplicity. For the possible worlds and individuals are eliminated in favour of propositions and properties. But there is also a loss in logical simplicity. For the quantifiers over propositions and properties are both second-order and intensional. It is not always appreciated that in modal logic the classical hierarchy of orders ramifies with respect to the extensional/intensional distinction. For example, at the second order there are extensional quantifiers over sets and intensional quantifiers over properties. Thus in the second modal language there is a double jump in logical complexity, one in order and the other from extension to intension.

This increase in complexity is not altogether necessary. For in the reverse translation, the intensional entities can be replaced by corresponding extensional entities, world-propositions by extensions and individual essences by singleton sets. Thus the intensional jump is avoided, though not the jump in logical order.

Let us consider the extensional substitute for individual essences in more detail. The basic idea is to replace the individual essence of a possible individual with the set whose sole member is that individual. Let us say, then, that X is a *singleton of* x if x is the sole member of X; and call X a *singleton* if it is possibly the singleton of some individual. In the reverse translation, the quantifier $\forall x$

will be replaced with a quantifier $\forall X$ over singletons and the predication $R'\mathbf{xy}w$ will be replaced with the formula 'Possibly there is an x and possibly there is a y such that p implies $x \varepsilon X$, $y \varepsilon Y$ and Rxy', where p is the world-proposition corresponding to w, and X and Y are the singletons corresponding to x and y respectively.

The extensional method for possible worlds is a little more complicated. In this case, the basic idea is to identify each possible world with the extension it confers on each of the predicates $R'_1, \ldots, R'_n$ in the classical language. (The method does not work for an infinity of predicates, although it is applicable to any particular formula.) Say that X is the *extension of* the binary predicate R if the true world-proposition p is such that necessarily for all x and necessarily for all y, p implies $\langle x, y\rangle \varepsilon X$ iff Rxy. There is a similar definition for predicates of arbitrary degree. (These definitions give the *outer* extension of a predicate, the set of possibles of which the predicate is true. It is more usual to consider the *inner* extension, the set of actuals of which the predicate is true.) Let an *extension-sequence* S be an n-tple $\langle X_1, \ldots, X_n\rangle$ such that it is possible that X_i is the extension of R_i for $i = 1, \ldots, n$. Then in the reverse translation, the quantifier $\forall w$ is replaced with the quantifier $\forall S$ over all extension sequences and 'p strictly implies $R_i xy$', in the formula that translates predications, is replaced with '$\langle x, y\rangle \varepsilon$ the i-th component X_i of S', where $S = \langle X_1, \ldots, X_n\rangle$ is the extension-sequence corresponding to w.

The method of substituting extensional entities does not avoid the need for existence assumptions. The use of singletons requires that singletons (and generally, sets) necessarily exist and that each individual determine a singleton. The second assumption follows from ordinary set-theoretic principles, but the first assumption can, and will, be questioned. The use of extensions requires that extensions necessarily exist and that each predicate determines an extension. In this case, both assumptions are questionable since ordinary set-theoretic principles only guarantee the existence of an inner extension (defined over actuals) and not of an outer extension (defined over possibles).

Prior used world-propositions for possible worlds and it is interesting to speculate on why he chose these intensional entities rather than some extensional counterpart. One reason may be that he was prepared to countenance non-nominal quantifiers, i.e. quantifiers whose variables do not occupy the same position as names. Given this stance, the simplest extensions of first-order or sentential modal logic are obtained by adding sentence or predicate quantifiers. And the most natural interpretation of these quantifiers —on logical and semantical grounds—is intensional. The logical ground is that only an intensional interpretation will guarantee the validity of Specification ($(\forall p)A(p) \supset A(B)$ in the case of sentences), and the semantical ground is that the free variable p should be interpreted in the same way as a sentence, viz. by assigning an intension. My own view is that all *primitive* quantifiers are nominal and that therefore all non-nominal quantifiers should be analysed away in terms of nominal quantifiers and perhaps other notions too. For me, then, there is not the above reason for preferring an intensional account of possible worlds.

4. *To be or not to be*

The previous reverse translations depended upon certain existence assumptions. There were the comprehension-type principles, that the appropriate world-propositions, individual essences, singletons and extensions exist. These principles were seen to be defensible, although a doubt was recorded for the case of extensions. Then there were the Barcan-type principles, that propositions, properties and sets necessarily exist. A more accurate formulation of these principles is that necessarily the respective entities necessarily exist. The principles, on the original formulation, would be vacuously true if there were no propositions etc. But this is a subtlety that need not concern us.

The time has come to criticize the second group of assumptions. Any such criticism will have far-reaching consequences, but our main concern will be with those consequences that affect the original programme of reduction. Consider first the case of sets. It is natural to suppose that sets are abstract and that therefore they necessarily exist. But take the set whose sole member is Socrates. Then surely it is necessary that it exists only if Socrates does. Since Socrates enjoys contingent existence, so does the set.

In the above argument, Socrates was only an example; any other contingent existent would have done instead. So what the argument establishes is that if some individual contingently exists then so does some set, i.e. that if all sets necessarily exist then so do all individuals. Since this argument is so critical, it may be worth setting down in formal detail. Let X range over sets and x over sets and individuals, and use E for the existence-predicate and ε for membership. The argument then goes as follows:

(1) $\forall X \Box EX$ (supposition)
(2) $\forall x \exists X(x \varepsilon X)$ (by set theory)
(3) $\forall x \exists X(x \varepsilon X \wedge \Box EX)$ (from 1 and 3 by modal logic)
(4) $\forall x \forall X(x \varepsilon X \supset \Box(EX \supset Ex))$ (assumption)
(5) $\forall x \forall X((x \varepsilon X \wedge \Box EX) \supset \Box Ex)$ (from 4 by modal logic)
(6) $\forall x \Box Ex$ (from 3 and 5 by classical logic).

The crucial assumption of the argument is (4), which says that if an object belongs to a set then necessarily the object exists if the set does. (4) follows by modal logic from two further assumptions:

(7) $\forall x \forall X(x \varepsilon X \supset \Box(EX \supset x \varepsilon X))$
(8) $\forall x \forall X(\Box(x \varepsilon X \wedge EX \supset Ex))$.

(7) is a rigidity condition; it says that a member of a set is necessarily a member—at least if the set exists. Thus the members of a set are preserved through possible worlds; they are essential to its identity. (8) says that an existent set cannot have non-existent members. It follows from the conception of sets as constructions; the set cannot be constructed if the members do not exist. The consequences of denying (8) are very strange. For example, Extensionality would fail since two sets might have the same existent members but different non-existent members. A condition would then no longer define a unique set and other means, presumably modal, would be required to define a set. Thus I am inclined to regard both (7) and (8) as correct essentialist principles concerning the nature of sets.

The admission of contingent sets opens up the field of modal set theory; it is no longer an uninteresting and trivial extension of the classical theory. This new field calls for a more careful and detailed treatment than I can give here, but let me consider the question of providing a criterion for the existence of sets. In possible worlds terms, a necessary and sufficient condition that a set exist in a world is that each of its members exist in that world. Given the Fundierungsaxiom, this is equivalent to saying that a set exists in a world iff each individual in its transitive closure exists in that world. (The transitive closure of a set consists of the members of the set, the members of the members, and so on.) Thus the existence of sets is ultimately traced back to the existence of individuals.

It is more difficult to state a criterion in modal terms. The necessity part of the above criterion is a strong form of (8), viz.

$$\Box \forall X \Box \forall x \Box (x \varepsilon X \wedge EX \supset Ex)$$

However, the sufficiency part cannot be similarly expressed since no first-order axioms will guarantee that the domain of sets in a world is the powerset of the domain of individuals. Suppose one assumes this and a strong form of rigidity, viz.:

$$\Box \forall x \Box \forall X \Box (x \varepsilon X \supset \Box (x \varepsilon X)).$$

Then the sufficiency condition is equivalent to the following modal form of Extensionality:

$$\Box \forall X \Box \forall Y (\Box \forall x (x \varepsilon X \equiv x \varepsilon Y) \supset X = Y).$$

Some of the above principles may strike the reader as unnecessarily subtle. But it is important to remember that the classical principles are, from the modal point of view, very weak. For example, the necessity of Extensionality merely says that no possible world contains distinct sets whose members, in that possible world, are the same. It says nothing about the membership of sets through possible worlds and so is compatible with sets being like boxes and changing their members from world to world. To state the identity conditions for sets in a satisfactorily strong metaphysical sense, one needs something like the above essentialist principles.

The objections to necessary existence, in the case of propositions, are more problematic. Reconsider the argument (1)–(6), but with X ranging over propositions, x ranging over objects, and 'ε' interpreted as 'occurs in'. Assumption (4) now says that a proposition cannot exist unless any object that occurs in it also exists. This may not be contentious. Unfortunately, the truth of (2)—that each object occurs in a proposition—is no longer clear.

First of all, the sense of 'occurs in' and of 'proposition' is obscure. If propositions are sentences and objects occur in sentences if their names do, then (4) is obviously correct but world-propositions will not, in general, exist and the reduction will not go through. In the context of modal logic and for the purposes of the reduction, it is most natural to say that propositions are identical if they are necessarily equivalent. One can then say that any object occurs in a proposition if the proposition cannot be expressed, in an ideal language, without reference to the object. An ideal language, here, is one that is not subject to finitary constraints; the alphabet can be of any infinite size and the logical combinations (conjunction, quantification, etc.) can be of any infinite length.

However, (2) is not true under this natural interpretation of propositional identity. It is tempting to suppose that any object x occurs in the proposition to the effect that x exists. But this is because the proposition is presented by means of a reference to the object. It may well be that the same proposition can be expressed without any reference to the object. Suppose, for example, that x is a broom. Then the proposition that the broom exists can be expressed by the sentence 'the brush and handle are attached' or its like, and so reference to the broom itself can be avoided.

There is, then, a fundamental difference between sets and propositions. For propositions, there is a distinction between superficial and deep structure. The objects that appear to be in the proposition may disappear, as it were, on analysis. But for sets, there is no such distinction. The structure of a set is not affected by the internal complexity of its members.

Despite this difference, it does seem reasonable to maintain that *some* contingent individual occurs in a proposition, i.e. that $-\Box Ex \wedge \exists X(x \varepsilon X)$ holds for some individual x under the earlier propositional interpretation of 'ε'. The argument (1)–(6), with '$\forall x$' dropped in (2)–(6), would then establish $\Box Ex$. This contradicts the above and so gives $-\forall X \Box EX$ by reduction. The existence of the individual x could be denied, but only by claiming that each proposition can be expressed without reference to any particular contingent individual. Although some metaphysical views imply this claim, it does seem preferable to make our reduction independent of such views.

A criterion of propositional existence can be given, although the formal details are beyond the scope of this paper. There is a linguistic criterion that is in analogy to the definition of occurrence;

it states that a proposition exists in a world if it can be expressed without any reference to possible individuals that do not exist in the world. There is also an equivalent non-linguistic criterion. Say that two worlds are *indistinguishable with respect to a* given world if they are identical but for the identity of their merely possible individuals, i.e. if there is an isomorphism from the domain of the one world on to the domain of the other that keeps fixed those individuals of the given world that exist in either of the other worlds. Then the criterion is that a proposition exists in a world iff the proposition is true in neither or both of any two worlds whenever those worlds are indistinguishable with respect to the given world; the proposition must not distinguish between indistinguishable worlds.

The reduction is principally concerned with the existence of world-propositions. Under certain circumstances it is possible to develop special criteria for the existence of world-propositions. Suppose, for example, that for any possible world there is an isomorphic world which has a disjoint domain of individuals. Thus in this set-up, the individuals are bare; they have no individual essence that can be expressed without reference to the given individual. Then under these circumstances, a world-proposition for v exists in w iff each individual that exists in v also exists in w. In particular, a world-proposition for v exists in the actual world only if each individual of v actually exists. In thinking of world-propositions, it is often helpful to bear the simple set-up above in mind.

There is no need to consider a separate argument for the contingent existence of some properties. For if the property of being identical to x exists then so does the proposition to the effect that x exists. Indeed, for any proposition there is a property whose existence-conditions are the same; if the proposition is q then the property can be the one that holds of 0 alone if q is true and holds of 1 alone if q is false. The existence criteria for properties can be developed in analogy to those for propositions.

Although some sets, propositions and properties contingently exist, there are large classes of these entities that necessarily exist. Call a set *pure* if it is built up from the null set (its transitive closure contains no non-sets) and say that a proposition or property is *purely general* if it can be expressed without reference to any individuals. Then the pure sets and the purely general propositions and properties all necessarily exist. The belief in the necessary existence of sets etc. may arise from an exclusive preoccupation with these special subclasses. One forgets that concrete objects may occur in the construction of an abstract object. In the case of propositions and properties, the preoccupation is further reinforced by the view that intensional entities are meanings that express contingent facts but exist independently of those facts.

5. *Trying again*

Let us consider how the reverse translations should be modified in the light of the previous criticisms. We consider first the reduction of possible individuals and then the reduction of possible worlds. The reductions of possible individuals to individual essences and to singletons, both possess unnecessary logical complexity. For the possibilist quantifier $\exists \mathrm{x}$ (= 'there is a possible individual x') can simply be replaced with $\Diamond \exists x$ (= 'possibly there is an actual individual x') or, alternatively, quantification over all possibles ($\forall \mathrm{x}$) can be replaced with necessary quantification over all actuals ($\Box \forall x$). Thus the external quantifier over possibles is converted into an internal quantifier over actuals. The translation of predications is also simplified, for $R'\mathrm{xy}w$ becomes 'q strictly implies Rxy', and similarly for the extensional account of possible worlds.

On the face of it, this method only provides a partial elimination of the possibilist quantifier. For $\exists \mathrm{x}B(\mathrm{x})$ says that some possible individual actually has B, while $\Diamond \exists xB(x)$ says that some possible individual has B in a world in which it exists; and so the two formulas may differ in truth-value if no possible individual that has $B(x)$ is actual. Say that a condition $B(x)$ is *strongly rigid* if it is true of a possible individual in every possible world whenever it is true of that individual in some possible world, i.e. if $\Box \forall x \Box (B(x) \supset \Box B(x))$ holds. Then it should be clear that the equivalence of $\exists \mathrm{x}$ and $\Diamond \exists x$ holds when both apply to a strongly rigid condition. But the reverse translations of ω-free classical formulas are always strongly rigid; this merely generalises the corresponding observation made for the case of a non-quantificational modal language. Therefore, within the context of the translation, the equivalence will always hold. The special properties of the reverse translation enable the partial elimination of $\exists \mathrm{x}$ by $\Diamond \exists x$ to be transformed into a complete elimination of possibilist quantifiers.

This method of elimination differs in a fundamental respect from the earlier reductions. They were *syntactically conservative* in the sense that the syntactic form of the classical sentences was preserved under translation. '$\forall x$' became '$\forall$ individual essences φ', and '$\forall w$' became '$\forall$ world-propositions q' or '$\forall$ extension-sequences Y'. On the other hand, the above method is *syntactically radical.* Syntactic form is not preserved, for the quantifier $\forall x$ is replaced with the hybrid form $\Box \forall x$.

The conservative/radical distinction has a quite general ontological significance. A conservative reduction enables one to identify certain objects in the reduced language with objects (the 'logical constructions') in the reducing language—individuals with essences, possible worlds with extension-sequences, reals with sets of rationals. On the other hand, a radical reduction does not, in general, enable such an identification to be made. Thus Russell's theory of descriptions does not give a construction that corresponds to each description and, similarly, rendering $\exists \mathbf{x}$ as $\Diamond \exists x$ does not enable one to say what possible individuals are.

Similar remarks will apply to the revised reductions for possible worlds: the quantifier $\forall w$ will always be replaced with an expression in which $\Box$ governs an actualist quantifier $\forall a$ (the range of 'a' will depend upon the reduction used). Possible objects have no exact identity for the modal actualist. But before, he was able to treat $\Diamond A$ as if it were existential and $\Diamond \exists x A$ as if it were doubly existential. This pretence can no longer be kept up. $\Diamond A$, when both translations are applied, will become a sentence of the form $\Diamond \exists a A$ and so the original modal form will be preserved. Once a possibility statement, then, always a possibility statement.

The same method of pseudo-quantifiers could be applied directly to the extensional and intensional accounts of possible individuals; *necessary* quantification over all singletons or all individual essences would then replace the non-modal quantifiers. However, this approach would merely re-create the earlier logical complexity. What motivated the original quantification over singletons and essences was the belief that these entities enjoyed necessary existence. But without this belief there is no longer any reason for making the logical climb to another order.

So let us reconsider the reduction of possible worlds, first on the extensional account and then on the intensional account. Before the reverse translation can be modified, it is necessary to be clearer about the behaviour of the membership relation ε. The question is: What truth-value, if any, should $a \varepsilon b$ have in a world in which one of a and b denotes an individual that does not exist? A more general question is: What truth-value, if any, should an atomic sentence have in a world in which one of the possible individuals named does not exist?

A general principle that helps resolve the above question is this: the truth-value of any formula, modal or non-modal, should not turn on the truth-value, if any, that is assigned to an atomic sentence (with a primitive predicate) in which one of the objects named does not exist. Say that a formula is *normal* if its truth-value remains fixed when the outer extensions of the predicates vary but the inner extensions remain the same. Then the principle states that each sentence should, on analysis, be equivalent to a normal sentence. This principle might be called the Indifference Principle.[1] It can be regarded as a consequence of Actualism. For it states that there are no genuine relations among non-existents or among non-existents and existents, that any relation which appears to discriminate among non-existents must be reducible to those that do not.

There are various ways in which this principle might be secured. One is to adopt the Falsehood Convention. This is the convention that any atomic sentence should be false in case of empty reference. In modal terms, this means that $\Box \forall_1 x_1 \ldots \Box \forall x_n(Rx_1 \ldots x_n \supset Ex_1 \wedge \ldots \wedge Ex_n)$ should be laid down as an axiom for each n-place primitive predicate R. Given these axioms, it is easy to show that any sentence is equivalent to a normal sentence. For if x is the sole variable in B, then $\Box B$ may be replaced with $\Box(Ex \supset B) \wedge (-Ex \supset B^f)$, where B^f is the result of replacing each atomic sentence containing x in B by $\perp$, the falsehood constant. There is a similar replacement in case B contains several variables. The result of the replacements is then an equivalent normal formula. One could, of course, equally well adopt a Truth Convention. But the Falsehood Convention seems to be more natural since one already understands most predicates as being false of non-existents. Another way of securing the Indifference Principle is to adopt the Gap Convention, that atomic sentences should be undefined in case of

[1] This principle is discussed in more detail in my 'Model Theory for Modal Logic—Part II', Submitted to the *Journal of Philosophical Logic*.

empty reference. One crucial difference between the Gap and Falsehood Convention is that the former does not admit a primitive existence-predicate, i.e. a predicate that is true of all existents and false of all non-existents in any given world. We shall, in the sequel, adopt the Falsehood Convention, although the Gap Convention will later be considered in a section on Q.

If the Falsehood Convention is adopted, then all predicates that do not conform to the convention must be defined in terms of predicates that do. This is relevant to the original question about the membership relation. For we may distinguish two senses of membership. There is *strong* membership—symbolised by ε'—that only holds of existents; and there is *weak* membership—symbolised by ε—that holds of x and y in a world whenever x belongs to y, regardless of whether x or y exist in the world. Now the Falsehood Convention requires that the strong should prevail. ε', rather than ε, should be taken as primitive, and ε should be defined in terms of ε' and perhaps other notions too, as long as they are false of non-existents. In fact, ε and ε' are interdefinable, as the following definitions show:

$$x \,\varepsilon\, y \equiv \Diamond(x \,\varepsilon'\, y)$$
$$x \,\varepsilon'\, y \equiv (x \,\varepsilon\, y \wedge Ex \wedge Ey)$$

(As definitions, these equivalences should hold for any possible individuals and any possible world, i.e. they should be true when prefixed with $\Box\forall x\Box\forall y\Box$. The same goes for the formulas below.) The compatibility of these definitions depends upon the truth of:

$$x \,\varepsilon'\, y \equiv (\Diamond(x \,\varepsilon'\, y) \wedge Ex \wedge Ey).$$

Given that ε' is only true of existents, this formula states that ε' is rigid; its inner extension in each possible world is the same. The correctness of the first definition, i.e. that ε is weak membership, depends upon the assumption that each member of a set can co-exist with the set. For without this assumption, a set would contain a member and yet in no possible world would both the set and the member exist.

In a similar way, one can distinguish between a strong and weak sense of identity. The strong sense—symbolised by $='$—only holds of existents; the weak sense—symbolised by $=$—holds of x and y in a world whenever x and y are identical, regardless of whether x or y exist in the world. Analogous interdefinability results hold: the definitions, compatibility and correctness conditions are obtained from the earlier ones by substituting $=$ for ε and $='$ for ε'. In this case, the correctness condition is obviously satisfied and the compatibility condition follows from the appropriate rigidity assumption, viz.:

$$x =' y \supset \Box(Ex \supset x =' y).$$

It is customary in modal logic to adopt $=$ rather than $='$ as primitive. This may be because the modal axioms then have a simpler formulation. Reflexivity, for example, requires no existential restriction. However, if the preceding considerations are correct it is $='$, not $=$, that is primitive. This difference is small, but not insignificant. If $='$ is primitive, then $=$ cannot be defined in non-modal terms since the use of $\Diamond$ or its cognates cannot be avoided; $=$, unlike $='$, would then appear to be an essentially modal notion.

The modified extensional reduction of possible worlds can now be given. The predication R'_ixy$w$ is replaced, as before, with '$\langle x, y\rangle \varepsilon$ the i-th component of S' or, alternatively, with '$\langle x, y\rangle \varepsilon X_i$'. '$\varepsilon$' is now the defined relation of weak membership. The quantifier $\forall w$ is replaced with $\Box\forall S$, where S ranges over extension-sequences, or, alternatively, $\forall wB$ is replaced with $\Box\forall X_1 \ldots \Box\forall X_n \Box((X_1$ is the extension of $R_1 \wedge \ldots \wedge X_n$ is the extension of $R_n) \supset B^*)$. In this way, one avoids the problem of putting all of one's extension-sequences in one worldly basket. For the quantifier $\Box\forall X_1 \ldots \Box\forall X_n$ will, in effect, run through all sequences of *possible* extensions.

The revisions do not depend upon the assumption that extensions necessarily exist. However, they still depend upon the assumption that the outer extension of a predicate in a world exists in some (possibly distinct) world. Now although each member of the outer extension exists in some world, there is no reason to suppose that all of the members can be found in a single world. Therefore, by our earlier criterion of set existence, there is no reason to suppose that the extension exists in any world. To take an extreme example, the extension in any world of $x = x$ is the set of all possible individuals, but in no world may all possible individuals be actual.

There are two solutions to this problem. One is to exclude from the language any predicate that is true of non-existents. The outer extension of each predicate will then coincide with its inner extension and the above difficulty will not arise. If the Falsehood Convention holds, then each primitive predicate will satisfy the restriction and so any language, when analysed, will submit to such a reduction. The other solution is to introduce a special quantifier PX over quasi-sets. Details will be given in the last part of §6.

Finally, let us reconsider the intensional reduction of possible worlds. It is again necessary to be clear about the primitives of the secondary modal language. We distinguish between sentences (such as 'Socrates is a philosopher') and names of propositions (such as 'the proposition that Socrates is a philosopher' and 'the proposition Socrates last expressed'). The semantical difference is between an expression having a proposition as intension and as extension. Before the propositional variables were sentential; they occupied the same position as sentences. But now the propositional variables are nominal; they occupy the same position as names of propositions.

Three predicates will be used to form sentences from names or variables for propositions: the existence- and truth-predicates and a predicate for strict implication. In conformity with the Falsehood Convention, the predicates for truth and strict implication will have the strong sense. Thus the strong truth-predicate—symbolised by T'—holds of a proposition ρ in world w iff ρ exists in w and ρ is true in w; and the strong implication predicate—symbolised by $\rightarrow'$—holds of propositions ρ and σ in world w iff ρ and σ exist in w and ρ implies σ. (If propositions are identified with sets of possible worlds, then ρ is true in w if $w \in \rho$ and ρ implies σ if $\rho \subseteq \sigma$.) Apart from general philosophical considerations, one drawback to using sentential variables is that they presuppose the weak rather than the strong truth-predicate. For under the ordinary semantics, the sentential variable p, like any sentence, will be true in a world independently of whether the proposition expressed by p exists.[1]

The behaviour of these predicates within modal contexts is quite

[1] At the end of my 'Propositional Quantifiers in Modal Logic' (*Theoria*, 36 (1970), pp. 336–46) sentential variables are combined with a variable domain of propositions. But I now consider this an incorrect approach.

distinctive. Many commonly accepted principles do not hold once the contingent existence of propositions is allowed. Let ξ be the operator 'the proposition that'; this ξ applies to a sentence to form a name of the proposition expressed by the sentence. Then the scheme $\Box(A \equiv T' \xi A)$, i.e. that A is necessarily equivalent to the proposition that A being true, is not, in general, correct. For let P be the sentence 'Socrates does not exist'. Then P is true in a world in which Socrates does not exist, but $T' \xi P$ is false since ξP does not exist. Indeed, in this case, the sentence $\Box(-P \equiv T' \xi P)$ is true. Strictly speaking, these conclusions hold only under certain assumptions about the existence of Socrates; they hold, for example, under the previously presented picture of bare individuals.

The rejection of this and related principles may appear strange. But it is perfectly natural once it is appreciated that $T' \xi A$ carries an existential commitment to the proposition that A. Since A need not carry this commitment, it can be true without $T' \xi A$ being true. If the commitment is made explicit, then the principles will stand. For example, the modified principle:

$$\Box(E \xi A \supset (T \xi A \equiv A))$$

is correct.

The existence-predicate enjoys peculiarities of its own. What is important to remember is that propositions may enjoy the same vicissitudes of existence as ordinary individuals. For example, take the principle of possible co-existence. This says that any two possible individuals both exist in some possible world or, to put it in modal terms, that $\Box \forall x \Box \forall y \Diamond (Ex \wedge Ey)$ holds. Now this principle is not especially plausible. But neither is the corresponding principle for propositions, that $\Box \forall \rho \Box \forall \sigma (E\rho \wedge E\sigma)$ holds. Indeed, on the simple picture of bare individuals, the two forms of the principle are equivalence, since then the proposition that x exists will exist iff x does.

If such existential failures are allowed, then many definitions of standard notions in the theory of propositions must be formulated with great care. Let us illustrate with the definitions of negation, disjunction, necessity and possibility. That ρ is the negation of σ—ρ neg′ σ—can be defined by $E\rho \wedge E\sigma \wedge (\forall \tau)(\rho \rightarrow' \tau \wedge \sigma \rightarrow' \tau \supset \Box T' \tau)$. But this defines strong negation. *Weak negation*—ρ neg σ—

can be defined by $\Diamond \rho$ neg$'$ σ. (The correctness of this definition depends upon the quite reasonable assumption that a proposition and its negation can always co-exist.) Given certain facts about neg, it is then quite easy to justify the use of a negation operation '*n*' on propositions. That ρ is the *disjunction* of σ and τ—ρ dis$'$ σ,τ—can be defined by $E\rho \wedge E\sigma \wedge E\tau \wedge \forall\mu((\sigma \rightarrow' \mu \wedge \tau \rightarrow' \mu) \equiv \rho \rightarrow' \mu)$. However, weak disjunction—ρ dis σ,τ—cannot be defined as $\Diamond(\rho \text{ dis}' \sigma,\tau)$, for there may be no world in which both σ and τ exist. Consequently, there is no justification of a disjunction-operation '*d*' in analogy to that for '*n*'. The *strong necessity predicate* on propositions—$N\rho$—can be defined by $\Box T'\rho$. Note that this is also the weak predicate, since necessary propositions necessarily exist. However, the *weak possibility predicate*—$P\rho$—cannot be defined by $\Diamond T'\rho$, for this definition requires that ρ both be true and exist in a possible world. Rather, the correct definition is $-\Box T'n(\rho)$.

These peculiarities would be avoided if implication or truth were taken in the weak sense, as free from an existential commitment to propositions. However, if the Falsehood Convention is accepted, these predicates must somehow be defined in terms of other notions. Our experience with identity and membership might suggest that it is an easy matter to define the weak predicates from the strong. But the success of the earlier definitions depended upon certain rigidity and existence conditions being satisfied. The truth-predicate is not rigid, it is a contingent predicate *par excellence*, and so the compatibility conditions will not even be satisfied. The implication-predicate is rigid (on existents), but the correctness conditions need not be satisfied. A proposition and one of its implications may fail to co-exist and $\Diamond(\rho \rightarrow' \sigma)$ would then be false of that pair of propositions. One might attempt to define $\rightarrow$ as necessity of the material implication, so that $(\rho \rightarrow \sigma) \equiv Nd(n(\rho),\sigma)$; but, as we have already noted, there is no justification for this use of '*d*'. It is, in fact, very difficult to define the weak notions in terms of the strong, and so we will follow the easier course of giving the reduction directly in terms of the strong predicates.

In giving the reverse translation, the definition of world-proposition must be modified to suit the nominal status of propositions. Thus ρ is a *world-proposition*—in symbols, $\underline{Q}\,\rho$—if $\Diamond(T'\rho \wedge \forall\sigma(T'\sigma \supset \rho \rightarrow' \sigma))$ holds. The reverse translation can now proceed as follows: the predication $R\mathrm{xy}w$ is replaced with

$\Box(\rho \supset Rxy)$; $\forall x$ is replaced with $\Box \forall x$; and $\forall w$ is replaced with $\Box \forall \rho$, where ρ is a quantifier over world-propositions.

That Q defines the correct notion requires a certain amount of justification. After all, might not a world contain no true proposition that distinguishes it from another world? This possibility can be ruled out on the basis of a metaphysical postulate. It says that each world can be described in terms of its actuals, i.e. that each world possesses a complete description which only makes reference to the individuals existing in that world. Thus, for a logical atomist, the description might be the conjunction of all atomic sentences of the form $Ra_1 \ldots a_n$, where $a_1, \ldots, a_n$ name actuals, and perhaps also of the statement $\forall x(x = a \vee x = b \vee \ldots)$, where the list $a, b, \ldots$ includes names for all actuals. For him, then, no two worlds could agree on the actuals they contain and on the relations that held among them and yet differ on the relations involving non-actuals. This postulate would seem to be a *sine qua non* of any reasonable actualist position. But once it is granted, the correctness of the definition of Q then follows. For the complete actualist description will express a proposition that both exists and is true in the given world.

Prior defines a world-proposition as one that is *maximally possible*;[1] it is possible and implies any proposition or its negation—in symbols, $P\rho \wedge \forall\sigma(\rho \rightarrow' \sigma \vee \rho \rightarrow' n(\sigma))$. The two definitions are equivalent if propositions necessarily exist. But not otherwise. Every world-proposition p is maximally possible or, at least, possibly maximally possible. For in the world in which ρ is true any proposition or its negation is true. However, not every maximally possible proposition need be a world-proposition. Pick a world v in which Socrates does not exist and form the conjunction of all propositions in v that are true of the actual world. Then the proposition expressed by the conjunction is maximally possible (in v), but it may not be a world-proposition. For although it implies that there is an individual which has the properties in fact possessed by Socrates, it may not imply that any particular individual has those properties. But any world-proposition must imply an instance of any existential proposition that it implies.

The use of the expression $\Box \forall \rho$ overcomes the problem of contingent propositional existence in the same way that the earlier use

[1] See Chapter 2 above, p. 43.

of $\Box\forall x$ and $\Box\forall X$ overcame the problems of contingent individual and set existence respectively. In effect, $\Box\forall\rho$ searches through each possible world for world-propositions. Since each world will, at least, contain its own world-proposition, all world-propositions will eventually be found.

It is interesting to consider the result of using $\forall\rho$ instead of $\Box\forall\rho$ in the original reduction. The use of $\Box\forall\rho$ was based on the equivalence $\Box A \equiv \Box\forall\rho(Q\rho \supset \Box(\rho \supset A))$. If the second '$\Box$' is dropped, we obtain the definition of a new operator $\boxdot$ as follows: $\boxdot A \equiv \forall\rho(Q\rho \supset \Box(\rho \supset A))$. Now the right-hand side is true in a given world iff A is true in all world-propositions that exist in the world. Say that v is *accessible from w* if the world-proposition for v exists in w. Then $\boxdot A$ is true in a world iff A is true in all accessible worlds.

This definition is of technical and philosophical interest. The technical interest is that it enables a weak modal operator with an accessibility semantics to be embedded in an extension of S5. The accessibility relation is reflexive, for each world contains its own world-proposition. But there is no reason why it should be transitive or symmetric. In the set-up of bare individuals, v will be accessible from w iff each individual in v is in w. In this case, the accessibility relation is also transitive and if certain reasonable assumptions are made about the distribution of individuals through possible worlds then $\boxdot$ will have exactly the logic of S4.

The philosophical interest is in relation to the thesis that necessity is truth in all possible worlds. It is often claimed that this thesis leads to the modal logic S5. But the above definition of $\boxdot$ is an instance of this thesis if $\Box(\rho \supset A)$ is interpreted as 'true in', and, as we have seen, the definition may lead to a logic weaker than S5. The explanation of this anomaly is that the thesis about necessity only holds when the domain of possible worlds is taken to be fixed. If the domain of possible worlds is allowed to vary from world to world, then a statement might be true in all actual possible-worlds, say, and yet not true in one of the possible-worlds of an actual possible-world. The Law $\Box A \supset \Box\Box A$ would then fail.

6. *Refinements and extensions*

The previous sections gave a modal reduction of a classical first-order language for possible worlds and individuals. The present section shows how the principles of the reduction can be applied to languages that are both weaker and stronger than the original classical language.

Identity for worlds. The classical language contains a relative identity predicate I' if the original modal language contains an identity predicate I. The ordinary identity predicate $=$ can be defined in terms of I' for $\mathbf{x} = \mathbf{y}$ iff $\exists\, wI'\mathbf{xy}w$. However, the classical language contains no identity predicate on possible worlds or, indeed, any predicate that is defined exclusively on possible worlds. Therefore it is natural to add such a predicate to the classical language even though it is not required for the classical translation.

In translating back from this enriched language, there are no difficulties when possible worlds are reduced to world-propositions. For then $w = v$ can be replaced with $\Box(p \equiv q)$ if sentential variables are used and with $\rho = \sigma$ (or $\Box(T'\rho \equiv T'\sigma)$) if propositional variables are used. However, there are difficulties when possible worlds are reduced to extensions. For, in this case, two worlds are identified if they confer the same extension on each predicate of the language, and so there is no way of saying that *distinct* worlds confer the same extension on the different predicate. If the language is such that this last possibility does not arise, then world identity does not call for special treatment. It may be defined in the classical language by conjoining formulas of the form $\forall\mathbf{x}\forall\mathbf{y}(R'\mathbf{xy}w \equiv R'\mathbf{xy}v)$.

This difference in the reductions shows that the intensional account is more basic than the extensional. The latter account works only when there are certain restrictions or conditions on the classical language. Once they are lifted, the intensional account must be used. This may be another reason why Prior did not consider an extensional account.

No sorts. The classical language is two-sorted. It contains a sort for possible worlds and a sort for possible individuals. Any many-sorted language can be reduced to a one-sorted language by collapsing all the sorts into a universal sort and then recovering the original sorts by means of sortal predicates. In the present case, the universal sort would be for possible objects (= possible worlds or possible individuals) and the sortal predicates would be for being a possible individual and for being a possible world.

It is sometimes possible to translate the de-sorted language back into the original many-sorted language. Suppose, for example, that crossing sorts always results in falsehood, i.e. that a predicate in the one-sorted language is false of objects $a_1, \ldots, a_n$ if one of them is not of the correct sort. Then a backward translation can be obtained by first replacing quantification $\forall x A(x)$ over the universal sort with $\forall x' A(x') \wedge \ldots \wedge \forall x^m A(x^m)$, where $x^1, \ldots, x^m$ are variables for each different sort, and then replacing predications $Ry_1 \ldots y_n$ with $\perp$, the falsehood constant, if the variables $y_1, \ldots, y_n$ are not properly sorted. In the present case, the backward translation is possible if possible worlds and possible individuals are distinct and if R' is true of possible objects $a_1, \ldots, a_n, a_{n+1}$ only when $a_1, \ldots, a_n$ are possible individuals and a_{n+1} is a possible world.

However, without special assumptions, the backward translation will not be possible. Consider the sentence that says something is of two sorts—for example, that some object is both a possible world and possible individual. Then, in the absence of special assumptions, no sentence can express this. The same difficulty arises if one adds to the sorted language a universal identity-predicate, i.e. one that does not discriminate between different sorts. For then the above sentence can be rendered as $\exists x^1 \exists x^2(x^1 = x^2)$.

All of the earlier reverse translations were for a two-sorted classical language. But suppose that the language is de-sorted or enriched with identity and that each possible world is counted as a possible individual. How, then, can the reverse translation be given? One method is to construct a first-order modal language in which the first-order variables range over both worlds and individuals. Thus the domain D_w associated with each world w includes both worlds and individuals proper. There is no need for the set of all worlds to be included in any domain, but one can at least require that $w \varepsilon D_w$, i.e. that each world belongs to its own domain. If earlier existence-conditions for world-propositions are carried over to worlds, then one could also require that $D_v \subseteq D_w$ implies $v \varepsilon D_w$, i.e. that any world contains a world constructed from its own individuals.

To this modal language is added an actuality predicate G. G is true of an object **x** in world w if (a) **x** is w and (b) **x** exist in w. Clause (b) is, of course, redundant if the condition $w \varepsilon D_w$ holds. Note that the classical correlate G' of G is the identity predicate on worlds.

Thus if one accepts the principle that R is a logical predicate if R' is, then G is a logical predicate. This, in a sense, justifies the introduction of G into the modal language. The reduction now proceeds as follows: R'xyz is replaced with $\Box(Gz \supset Rxy)$ and Wx = 'x is a world' with $\Diamond Gx$; ∀x is replaced with $\Box\,\forall x$, as before; and $A(\omega)$ is replaced with $\exists x(Gx \wedge A^*(x))$, where $A^*(x)$ is the modal formula corresponding to $A(x)$.

The previous translations all involved a jump in expressive power: the classical language was stronger than the primary modal language; and the secondary modal language was, in its turn, stronger than the classical language. However, in this case, the modal and classical languages are on an expressive par. For a classical translation of the new modal language merely returns the original one-sorted classical language.

However, from the point of view of modal actualism, the new language is not above suspicion. It is not that it offends against actualism; the domain of each possible world need only contain the existing individuals and worlds. Rather, it goes against the spirit of modalism. Possible worlds quantifiers and modal operators are usually regarded as two competing means of expressing modal truths. The new language is a curious hybrid in which both means of expression are combined. Since the modalist does not need both modal operators and possible worlds, it is only natural for him to eliminate the latter along lines already indicated. The classical possibilist, too, may attempt to eliminate the modal operators in favour of possible worlds; but for him there may be special difficulties, since not all of the possible worlds may exist in the actual world.

Rigidity operators. Our main concern, so far, has been to translate a classical language of possible objects into a modal language. However, the same principles of translation may be used in translating one modal language into another. We now show that certain non-standard operators can be defined in terms of the standard ones. What distinguishes the non-standard operators is that a reference to the actual world is involved in their truth-conditions.

First we consider the case in which the operator $\downarrow$ for 'it is actually the case that' is introduced into the language.[1] The semantics for

[1] The temporal counterpart 'Now' is considered by H. Kamp in 'Formal Properties of "Now",' *Theoria*, 37 (1971), pp. 227–73.

this operator is that $\downarrow A$ is true at a possible world if A is true in the actual world. Thus if $A^*(w)$ corresponds to A, then $A^*(\omega)$ corresponds to $\downarrow A$. This operator increases the expressive power of ordinary first-order modal logic. For example, the sentence $\Diamond \forall x - \downarrow Ex$, saying that it is possible that all actual individuals do not exist, cannot be said without the use of $\downarrow$.

The language with $\downarrow$ can be translated into the classical language and so it can be translated into the secondary modal language. A direct translation can also be given. Thus if A is of the form $A(\downarrow B)$ it may be translated as $\exists p(p \wedge Q p \wedge A(\Box(p \supset B)))$ or as $\exists \rho(T'\rho \wedge Q \rho \wedge A(\Box(T'\rho \rightarrow B)))$ if nominal quantifiers for propositions are used. Similarly, if A contains several formulas of the form $\downarrow B$. In effect, $\downarrow B$ is replaced by 'The true world-proposition strictly implies B', where the description is given widest scope. The translation could also be based upon the extensional account of possible worlds. For example, the sentence $\Diamond \forall x(- \downarrow Ex)$ would then become $\exists X(\forall x(x \varepsilon X) \wedge \Diamond \forall x(-x \varepsilon X))$.

The behaviour of $\downarrow$ may, in some ways, appear anomalous. The formula $A \supset \Box \downarrow A$ is valid, but the necessary truth of $\downarrow A$ does not follow from the truth of A. However, this behaviour is merely the result of scope ambiguity. The conditional formula, upon translation, becomes $\exists p(p \wedge Q p \wedge (A \supset \Box\Box(p \supset A)))$; and the validity of this formula does not imply that $\exists p(p \wedge Q p \wedge \Box(p \supset A))$ is necessarily true if A is true. Admittedly, it is odd to regard a connective as being subject to scope ambiguity, but the use of the description 'the true world-proposition' shows how this ambiguity can be reduced to the more familiar case of descriptions.

Kaplan[1] so interprets $\downarrow$ that $\downarrow A$ is necessarily true if A is true. This interpretation has some extreme consequences: it leads, for example, to a complete breakdown in the correlation between *a priori* and necessary truths. What the above analysis shows is that the logical behaviour of $\downarrow$ can be respected and yet the extreme consequences avoided.

Vlach[2] has introduced a companion † for $\downarrow$. The tense-logical readings for † and $\downarrow$ are 'once' and 'now' (or 'then'). †, however, has no modal reading. The semantics for † and $\downarrow$ can be explained

[1] 'The Logic of Demonstratives' (unpublished).

[2] 'Now and Then', presented to the Australian Association for Logic, August 1970.

in the following way. Usually, when evaluating a formula at a world, one has to evaluate its subformulas at different worlds. But once this is done, the reference to the original world is lost. Now † allows one to keep or 'store' a reference to the world of evaluation, while ↓ enables one to pick up this reference. Thus † and ↓ work in pairs, with ↓ picking up the reference fixed by †. In evaluating a formula, one must consequently keep track of two worlds: the 'floating' world at which ordinary formulas are evaluated and the (temporarily) 'fixed' point. The rules for † and ↓ are then: $\dagger A$ is true when the floating world is w if A is true when the fixed world is w (keeping the floating world the same); and $\downarrow A$ is true when the fixed world is w if A is true when the floating world is w (keeping the fixed world the same).

The above rules permit a translation of the language with † and ↓ into the classical language and thereby into the secondary modal language as well. Rather than give details, let us simply give the direct translation of † and ↓ into the secondary modal language. If C is a formula $\dagger A(\downarrow B)$ that only contains the two designated occurrences of † and ↓, then its translation is $\exists p(p \wedge Qp \wedge A(\Box(p \supset B))$. This replacement can take place when C itself is part of a broader formula. Thus the effect of † is merely to limit the scope of the world-description implicit in ↓; † is, in terms of the preceding discussion, a marker of scope. Any adjacent pairs of † and ↓ can be eliminated in this way. If any occurrences of ↓ remain, they may be singly eliminated as above. (In effect, a sufficient number of occurrences of † are placed at the beginning of the formula.) If any occurrences of † remain, they may be dropped since the references secured by † are never picked up by a subsequent occurrence of ↓.

Possibilist quantifiers. We now consider a modal language in which the quantifiers range over all possible individuals. Such a language is in conformity with modalism, but not with actualism, and so the possibilist quantifiers must be removed. We have already observed that $\Diamond \exists x A(x)$ is not, in general, a correct translation of $\exists \mathbf{x} A(\mathbf{x})$. However, the translation may be repaired with † and ↓; $\exists \mathbf{x} A(\mathbf{x})$ is equivalent to $\dagger \Diamond \exists x \downarrow A(x)$. For when we come to evaluate $\downarrow A(x)$, ↓ takes us back to the point at which we evaluated $\dagger \Diamond \exists x \downarrow A(x)$. Since † and ↓ can be eliminated, this gives a direct translation: $\exists x A(x)$ goes into $\exists p(p \wedge Qp \wedge \Diamond \exists x \Box(p \supset A(x)))$. A similar result could also be obtained, of course, by looking at the classical translation of the language with possibilist quantifiers.

It is sometimes claimed that certain English sentences require possibles for their analysis and that this constitutes an argument for the possibilist's position.[1] Such arguments are quite naive, since they presuppose that the logical form of the sentences will closely follow their grammatical form. It is true that some uses of the possibilist quantifier are not eliminable within a first-order actualist language. The formula $\Box \exists x Px$ provides a particularly simple example. However, as the preceding translation shows, all of the uses are eliminable once propositional quantifiers are added to the actualist language.

Higher-order logic. Prior's reduction of possibilist talk was, in one respect, over-modest. He was content to give a modal reduction of that language required for the classical translation of modal discourse. But the resulting modal language could also be translated into a classical language which, in its turn, could be given a modal reduction, and so on *ad infinitum*. In general, the problem remains of reducing a higher-order classical language, with quantifiers over sets of possible worlds, sets of possible individuals, sets of such sets and so on.

In case propositions, properties and sets enjoy necessary existence, this problem has a simple solution. For possible worlds and individuals were identified with objects of a certain sort and so sets of the former entities could be identified with sets of the latter entities; for example, sets of possible worlds could be identified with sets of world-propositions. But without necessary existence, the reverse translation is syntactically radical and so there is no such automatic extension to higher-order logic. For example, what corresponds to the possible individuals quantifier is the modalised quantifier $\Box \forall x$, where x ranges over actual individuals. But what is to correspond to the quantifier over sets of possible individuals? Should it be $\Box \forall X$, where X ranges over (actual) sets of individuals? But then one would have to assume that each set of possible individuals exists in some possible world or, in terms of our criterion for set existence, that each set of possible individuals is included in the domain of some world. And this is not a reasonable assumption.

[1] Cf. *Past, Present and Future*, pp. 149–51.

There may be better solutions. For example, sets of possible individuals may be replaced with properties and membership regarded as possible possession of the property. There will, however, be no general solution within the type of modal language so far considered. Suppose that all possible worlds are isomorphic and that their domains are finite but disjoint. Then in each possible world the domains of sets, properties, propositions, etc., will be extremely limited; and it would appear to be impossible to say various things—such as that there are an infinite number of possible individuals.

There are, as with the analogous problem over the extensional account of possible worlds, two responses to this difficulty. First, one might rule out those situations that give rise to this limitation in expressive power. It may be that no reasonable metaphysical view would tolerate the type of situation described above, but all the same it would be preferable to free the reduction from all but the weakest metaphysical presuppositions. The other response is that the higher-order extensions of a modal language have been too narrowly conceived and that some distinctively modal types of variable-binding operator need to be introduced. More specifically, we need a variable-binding operator P on quasi-classes $\otimes$ and a relation-symbol 'In' that holds between individuals and quasi-classes. $(P\otimes)B$ is true in a world iff for some *arbitrary* class of possible individuals, B is true in each world; and 'x In $\otimes$' is true in a world when the object assigned to x belongs to the class assigned to $\otimes$.

Now the natural objection to this move is that $P\otimes$ is merely a possibilist quantifier over all sets and that therefore the new language is not in conformity with actualist principles. But I am not sure that we need accept this objection. Let B^+ be the result of replacing 'x In $\otimes$' in B by the infinite disjunction $x = x_1 \vee x = x_2 \vee \ldots$ for arbitrarily many variables $x_1, x_2, \ldots$ Then $(P\otimes)B$ can be regarded as an abbreviation for the infinitary but actualist formula $\Diamond \exists x_1 \Diamond \exists x_2 \ldots \Box B^+$. For this latter formula says that are possible individuals $\mathbf{x}_1, \mathbf{x}_2, \ldots$ (not necessarily all in one world) such that B^+ holds, and $x = x_1 \vee x = x_2 \vee \ldots$ holds of $\mathbf{x}$ and $\mathbf{x}_1, \mathbf{x}_2, \ldots$ if 'x In $\otimes$' holds of $\mathbf{x}$ and the class $\{\mathbf{x}_1, \mathbf{x}_2, \ldots\}$. In the same way, one might regard the quantifier $\exists X$ over classes of individuals as an abbreviation for $\exists x_1, \exists x_2, \ldots$, with $x \varepsilon X$ replacing

$x = x_1 \vee x = x_2 \ldots$ In both these cases, a new notation is introduced in order to give finitary expression to a formula that is infinitary but subject to certain constraints. In the first case, the constraint is that the quantifiers be actualist and, in the second case, it is that the quantifiers be first-order.

The notation $\Diamond \exists x_1 \Diamond \exists x_2 \ldots \Box$ of the expansion is, I think, very natural within the context of an infinitary modal language with quantifiers. For suppose that we wish to give a complete description of all the modal facts, i.e. of the universe of possible worlds as opposed to any of its members. Then it is natural for the modal actualist to proceed in the following way. First, he gives a complete description $B_1, B_2, \ldots$ of each world as if he had names for all possible individuals. Second, he replaces those names by variables $x_1, x_2, \ldots$ to give the formulas $C_1, C_2, \ldots$ Finally, he gives the formula $A = \Diamond \exists x_1 \Diamond \exists x_2 \ldots \Box(x_1 \neq x_2\, x_1 \neq x_3 \wedge \ldots \wedge \Diamond C_1 \wedge \Diamond C_2 \wedge \ldots \Box(C_1 \vee C_2 \vee \ldots))$ as the complete description of the modal universe. The use of names in the first stage is, of course, a mere pretence used in defining the final formula A. For the modal actualist, there may be no specific individuals $a_1, a_2, \ldots$ in virtue of which the formula A is true.

The use of the notation $P\otimes$ (or $\forall X$) is justified in terms of the infinitary expansion, and it is necessary that any extension of the notation be justified in terms of a corresponding extension of the expansion. For example, it is easy to extend the notation to quasi-classes of ordered pairs (whose components can co-exist), but there is no basis for allowing $\otimes$ to appear to the left of 'In' (or allowing X to appear to the left of 'ε'). Thus the uses of $P\otimes$ are severely limited. Intuitively speaking, one should always be able to regard a quasi-class as distributing over all possible worlds, as picking out possible individuals from each of the worlds in turn.

As long as this limited use of quasi-classes is accepted, one can resolve the above difficulty of expressing that there are an infinite number of possible individuals. For one may say that there is a quasi-set $\otimes$ of pairs$\langle n, \mathrm{x}\rangle$, n a natural number and x a possible individual, such that it defines a one–one function on the natural numbers. One can also solve the problem in §5 over the extensional account of possible worlds. For the quantifiers over extensions can be replaced with quantifiers over quasi-sets.

Indeed, it would appear that quasi-sets provide a general reduction of higher-order classical languages. For we may say that there exists a quasi-set $\otimes$ of pairs $\langle \alpha, t \rangle$, α an ordinal and t a world-proposition or individual, that is a one–one function from a class of ordinals on to the class of world-propositions and individuals. In terms of this one–one correspondence, one may define relations R^* on ordinals that are isomorphic to the relations $\Box(T'\rho \supset Rxy)$ between world-propositions and individuals. One then has, in any possible world, a structure that is isomorphic to the one provided by the classical language, and so one may use any resources from the classical language in describing this structure.

7. *Q-ish difficulties*

We single out for separate treatment the problem of reduction for Prior's system Q[1]. The system Q is an actualist modal logic. What distinguishes it from other such logics is its account of sentences which contain names for individuals not existing in a given world. All such sentences are said to be undefined or truth-valueless. This might be called the Gap Convention, in analogy to the Falsehood Convention which states that atomic sentences are false in case of empty reference.

The Gap Convention might be broken down into two parts: first, that atomic sentences are undefined in case of empty reference; and second, that gaps are preserved under the logical operations, be they truth-functional, modal or quantificational. It would then be possible to accept the first part of the convention but not the second (the second part is vacuous, of course, if there are no gaps). For example, gaps could be treated by the method of supervaluations or by some other set of three-valued truth-tables. However, such a differential approach goes against the whole spirit of Q. An empty name in any sentence, be it atomic or not, is a source of gaps. For in a world in which reference fails, the sentence is not 'statable', it states no fact of the world, and so cannot be given a truth-value. Segerberg introduces an operator T (taking a gap into false) in proving completeness for Q-type system.[2] But if the above remarks

[1] *Time and Modality.*

[2] 'Some Modal Logics Based on a Three-Valued Logic', *Theoria*, 36 (1970), pp. 301–21.

are correct, such operators have no place in a proper presentation of the system Q.

The system Q is, in many respects, unnatural. Because of gaps, many standard modal principles are only valid under statability restrictions. Again, gaps create the need for distinguishing between a strong and weak sense of the modal operators. S is strongly necessary ($\boxed{S}$) if it is always true and weakly necessary ($\boxed{W}$) if never false; and similarly, S is strongly possible (Ⓢ) if it is sometimes true and weakly possible ($\boxed{W}$) if sometimes not false. Perhaps odder than these two features are certain anomalies in what can be expressed and how. One would naturally express 'Possibly I do not exist' with the sentence $\Diamond - Ea$, where $\Diamond$ has the strong sense. But, in Q, E is undefined for non-existents and so Ⓢ $- Ea$ is false. What $\Diamond - Ea$ expresses in our earlier modal language would be expressed in Q by $\boxed{W} - Ea$. However, this is a kind of accident. For what is earlier expressed by $\Diamond(-Ea \wedge -Eb)$—e.g. 'Possibly I and you do not exist'—cannot be expressed in Q at all. Only in higher-order extensions of Q could this sentence be expressed. In first-order Q, there are definite limitations on what uses of E can be captured.

Prior was well aware of the first two oddities, though perhaps not the third. All the same, Prior was firmly convinced, on philosophical grounds, that Q was the 'correct' modal system. I do not think his arguments for this claim can be sustained, but a full discussion of the issue would take us too far afield. Let me merely discuss the issue insofar as it relates to the preceding account of propositional existence.

The question is whether a predicate can be true (or false) of non-existents. In particular, we may ask whether the existence-predicate can be false of a non-existent, or, to put it in modal terms, whether $\exists x \Diamond - Ex$ can be true for the strong sense of $\Diamond$. For the sake of an example, let S be the sentence 'Socrates does not exist'. Then Prior would argue as follows:[1]

(1) $\Diamond S$ is true only if the proposition that S is true in some possible world ($\Diamond S \supset \Diamond T' \xi S$).
(2) The proposition ξS is true in a world only if it exists in that world ($\Box(T' \xi S \supset E \xi S)$).

[1] *Past, Present and Future*, pp. 149–51.

(3) The proposition ξS exists in a world only if Socrates exists in that world ($\Box(E\,\xi\, S \supset Ea)$, for $a =$ 'Socrates').

(4) The proposition ξS is true in a world only if Socrates does not exist in that world ($\Box(T'\,\xi\, S \supset -Ea)$).

(5) From (1)–(4) it follows that $\Diamond S$ is false ($-\Diamond S$).

Now (4) is certainly correct, and (3) we are prepared to accept. (2) is also correct, at least for a strong sense of truth. (1), however, is not acceptable. For, as we have already remarked, $\Box(S \supset T'\,\xi\, S)$ is not true, and so there is no reason why the truth of $\Diamond S$ should imply the possible truth of ξS.[1]

The matter cannot, however, rest there. For something like (1) must be correct if the possible worlds semantics is to be applicable to modal languages. One solution to this problem is to adopt the principle that $\Diamond S$ is true if the sentence S is true in some possible world, and yet not require that when a sentence is true in a world then so is the proposition expressed. A related solution is to adopt (1) but for a weak sense of 'true in' or 'true' (T, not T'), so that a proposition can be true in a world without the proposition existing in that world. If either solution is adopted then the relevant senses of 'true in' must be made out. Now that S is true in a world might be expressed in terms of the truth of $\Box(T'\rho \supset S)$, where ρ is a world-proposition. The Q-theorist could, of course, rejoin that the truth of $\boxed{S}(T'\rho \supset S)$ implies the necessary existence of ξS. But as an argument *for* Q, this is circular. The most that the Q-theorist can argue is that Q is self-sustaining, that the relevant sense of 'true in' cannot be made out once a Q-ish language is adopted. But this is not to argue for Q on the basis of a neutral position in which only actualism and the legitimacy of the possible worlds semantics are taken for granted.

Even though I reject the arguments for Q, I do think that the system embodies correct principles and distinctions. For, first, Q is in conformity with the Indifference Principle, the principle that the truth-value of a formula should not turn upon relations involving non-existents. In Q, the principle is secured with the Gap Convention; whereas I should prefer to reconcile the principle with the absence of gaps. And, second, the distinctions arising from gaps can be re-expressed in terms of propositional existence. For

[1] This question is briefly discussed in *Papers on Time and Tense*, pp. 128–31.

example, $\boxed{\text{s}}A$ is, in our terms, $\Box T' \xi A$, whereas $\boxed{\text{w}}A$ is $\Box(E \xi A \supset T' \xi A)$.[1]

Let us now turn to the translations from and to Q. In the first direction, there are various classical translations and consequently various classical languages that might be used. First, the classical language might contain gaps that correspond to the gaps in the original modal language. Thus Rxy would go into $R'\mathrm{xy}w$, where the second formula is truth-valueless whenever one of the individuals assigned to $\mathbf{x}$ or $\mathbf{y}$ did not belong to the world assigned to w; v and $-$ would represent themselves, though gaps would need to be preserved in the classical language too; $\boxed{\text{s}}B$ would go into $\forall w TB^* \wedge SB^*$ and $\boxed{\text{w}}B$ into $\forall w - T - B^* \wedge SB^*$, where T is the truth-operator (true when the sentence is and false otherwise) and where SA abbreviates $A \mathrm{v} - A$; and, finally, $\forall xB$ would go into $\forall \mathrm{x}(TE'\mathrm{x}w \supset TB^*) \wedge SB^*$. Second, to each modal formula A there might correspond two conditions $A^t(w)$ and $A^s(w)$ that give the truth and statability conditions of A respectively. $A^s(w)$ would be the conjunction of the formulas $E'\mathrm{x}w$ for $\mathbf{x}$ free in A and $R^s w$ (meaning R is statable in w) for R a predicate of A. As for $A^t(w)$, Rxy would go into $R^t\mathrm{xy}w$, where this is now a bivalent formula; $(-B)^t$ into $-(B^t)$; $(B\mathrm{v}C)$ into $(B)^t\mathrm{v}(C)^t$; $\boxed{\text{s}}B$ into $\forall w B^t(w)$; $\boxed{\text{w}}B$ into $\forall w(B^s(w) \supset B^t(w))$; and, finally, $\forall xB$ into $\forall \mathrm{x}(E'\mathrm{x}w \supset B^t)$. Finally, Q could be translated into *our* first-order language and thereby into the standard classical language. Everything but the modal operators would be untouched. If A contained only the variables x, y, z, then $\boxed{\text{s}}A$ would go into $\Box(Ex \wedge Ey \wedge Ez \wedge A)$ and $\boxed{\text{w}}A$ would go into $\Box(Ex \wedge Ey \wedge Ez \supset A)$. In this translation, unlike the others, it would need to be assumed that no variables (or, generally, names and variables) were implicit in the simple predicates. Thus a formula would be statable just when the explicitly named individuals existed.

The secondary modal language is obtained from the original one by adding sentential variables $\forall p$. However, propositions can no longer be regarded as sets of possible worlds; for a proposition may be true, false or non-existent (unstatable) in any given world. Instead, a proposition may be regarded as a *partial* function from the set of possible worlds into the two truth-values, True and False.

[1] Cf. the last paragraph of 'The Possibly Time and the Possible', *Mind*, 78 (1969), pp. 481–92.

The quantifier $\forall p$ will then range, in each world, over all propositions that are defined for that world. There is, in Q, no reason for letting the domain of propositions (i.e. partial functions) vary from world to world. For there are no propositions, like our earlier proposition that Socrates does not exist, which are true or false in a world in which they do not exist. The contingency of propositional existence is already captured, as it were, in the possible worlds functions being partial.

In Q, there are two kinds of world propositions. On the one hand, there are those that are true in one world and false in all others. These are definable by $\circledS(p \wedge \forall q((q \wedge \boxed{\text{S}}(q \supset q)) \supset \boxed{\text{S}}(p \supset q)))$ as long as they exist; but there is no reason, in general, to suppose that such world-propositions exist. On the other hand, there are the propositions that are true in one world and not true in all others. These propositions exist, but are not definable. The formula $\circledS(p \wedge \forall q(q \supset \boxed{\text{W}}(p \supset q)))$ defines them under the assumption that any two worlds are sharply distinguishable, i.e. that there is a proposition true in one of them and false in the other. However, this assumption is questionable (consider two isomorphic worlds with disjoint domains of individuals). There seems to be no other definition that would avoid this difficulty; and so let us suppose that some assumption or new primitive enables the second world-propositions to be defined by a formula $Q\,p$. The need for these props might be taken as an objection, albeit of a technical and recondite sort, against the system Q.

A reverse translation can now be given for the various classical languages, although we shall only consider the last. First, $R'\mathrm{xy}w$ is replaced with $\circledS(p \wedge Rxy)$ and $\forall wA$ with $\boxed{\text{W}}\,\forall p(Q\,p \supset A^*)$. Note that $\boxed{\text{W}}(p \supset Rxy)$ cannot be used for atomic sentences since it is true when Rxy is unstatable in the given world. Finally, if A is a formula whose sole free variables are x and y and A is the result of replacing each atomic formula containing a free occurrence of y in A^* by $\perp$, then $\forall xA$ is replaced with $\boxed{\text{W}}\,\forall xA^* \wedge \boxed{\text{W}}\,\forall p((Q\,p \wedge -\circledS(p \wedge Ey)) \supset \circledS(p \wedge \forall xA))$; there is a similar replacement when more variables occur freely in A. The first conjunct only says that A is true of all individuals x that co-exist with y, and this explains the need for another conjunct. The replacement A^e in this conjunct is justified as long as the classical language conforms to the Falsehood Convention. The translations of $\forall w$ and $\forall x$ are only correct under the assumption that for any finite set of individuals and world-proposition there is a world in which the individuals exist and the proposition is statable. Again, this assumption appears to be both questionable and unavoidable.

It is worth noting that the translation only uses $\boxed{\mathrm{w}}$ and its companion operator $\circledS = -\boxed{\mathrm{w}}-$. This shows that $\boxed{\mathrm{s}}$ (or $\boxed{\mathrm{w}}$) must be definable in terms of $\boxed{\mathrm{w}}$ and sentential quantifiers. Suppose that A contains the free variable x alone. Then a direct translation of $\boxed{\mathrm{s}}A$ is $\boxed{\mathrm{w}}A \wedge \boxed{\mathrm{w}}\,\forall p(Q\,p \supset \circledS(p \wedge Ex))$. A similar translation can be given in case A contains several variables.

8. *Problems with time*

The preceding sections have been entirely occupied with modal logic. Many of the considerations apply, *mutatis mutandis*, to tense logic and so a separate detailed account is not required. What we shall do is consider the special problems when tense logic is substituted for modal logic and when it is combined with modal logic.

The original tense-logical language is like the original modal language, but with the two tensed operators G (always will be) and H (always has been) in place of the single modal operator $\Box$. The corresponding classical language is still two-sorted, but now one sort is for individuals—past, present or future—and the other sort is for instants of time. We use x, y, z, . . . as variables for the first sort and t, u, v, . . . as variables for the second. The language contains an earlier-later relation $<$ in addition to the predicates R' with an extra argument for instants. The classical translation is then as follows: Rxy goes into $R'\mathrm{xy}t$; the connectives v and $-$ represent themselves; GA goes into $\forall u(t < u \supset A^*(u))$ and HB into $\forall u(u < t \supset B^*(u))$; and $\forall xB$ goes into $\forall x(E'xt \supset B^*(t))$.

Let us note that there are now two interpretations of E (or E'). On the first interpretation, only present individuals exist. The domains of individuals at different instants could then vary in size, although one might suppose that individuals don't exist after ceasing to exist, i.e. that $D_t \cap D_v \subseteq D_u$ for $t < u$ and $u < v$. On the second interpretation, past and present individuals exist. The domains of individuals are then cumulative, i.e. $D_t \subseteq D_u$ for $t < u$. E could, of course, be used for present existents alone and then the cumulative interpretation of the quantifiers would require the translation $\forall x(\exists\, u(u < t \wedge E'xu) \supset B^*(t))$. I myself find the second form of tense-logical Actualism more plausible than the first. For certain causal facts seem to involve a 'genuine' relation between individuals at different instants.

For the reverse translation, it is necessary to define instant-propositions. Let us use LA for 'it is always the case that A' and MA $(= -L-A)$ for 'it is sometimes the case that A'. If $<$ is linear, then LA can be defined as $HA \wedge A \wedge GA$. If time is not linear, then a new definition or a new primitive may be required. That p is an *instant-proposition*—in symbols, Ip—can now be defined as $M(p \wedge \forall q(q \supset L(p \supset q)))$. If time is linear, then Ip can be defined without quantifiers as $M(p \wedge G - p \wedge H - p)$. However, this is not of much help since quantifiers will eventually need to be introduced. The reverse translation is then as follows: $R'\mathbf{xy}t$ is replaced with $L(p \supset Rxy)$ and $t < u$ with $L(p \supset Fq)$; $\forall t$ is replaced with the quantifier $\forall p$ over instant-propositions; and $\forall \mathbf{x}$ is replaced with $L\,\forall x$.

The only new feature of the account is the definition of $<$ as $L(p \supset Fq)$. An equally good definition is $L(q \supset Pp)$. These definitions are interesting in that they are the only place in the translation where one of the uni-directional operators P or F is used in preference to the bi-directional operator L. Since each definition uses only one of the operators P or F, it should be possible, via the two translations, to define either operator in terms of the other and L. For example, a direct definition of FA is $\exists p(p \wedge Qp \wedge M(A \wedge Pp))$: A is future if at some time it is the case when the present is past. These definitions may suit those who are inclined to regard one of the tense-logical operators as more basic than the other, though it is hard to justify the choice of L as primitive.

This account will raise the same questions of existence for instant-propositions as previously arose for world-propositions. First, there will be the question of showing that instant-propositions always exist. It is tempting to follow the earlier arguments and say that an instant-proposition is the conjunction of all (tensed) propositions true at a given instant. However, the nature of the propositions that need to be chosen will be different in the cases of time and modality. To describe a possible world, one need only use propositions that are expressed by non-modal sentences, the logical modalities are simply not required in describing empirical reality. However, to describe an instant one may need to use propositions that are expressed by sentences containing tense-logical operators. One may need to say that such and such *will* happen or *has* happened. There do, indeed, seem to be cases in which no use of the tense-logical operators is required—e.g., 'Jesus Christ is born'. But, first, it is not clear that this sentence picks out a single instant and, second, the analysis of 'born' may well require the use of tense-logical operators. It follows that Prior's reduction of instants may be substantially different from one in terms of events, for future and past-tensed sentences are not usually regarded as describing present events.

Secondly, there is the question of whether instant-propositions (or tensed propositions, generally) always exist. Just as the modal actualist can claim that propositions about Socrates do not exist in worlds in which Socrates does not exist, so his tense-logical counterpart can claim that (tensed) propositions about Socrates do not exist at instants at which Socrates does not and, perhaps, has existed. The solutions for the two cases are also similar, and so I shall not give separate details. Let us merely note that the cumulative interpretation of the quantifiers provides a special bonus; for if two objects exist at different instants they also exist at the same instant, viz. the later of the two instants, and so there will be simple definitions of the weak notions of §5 above in terms of the strong.

The tense-logical case also raises questions of its own. Can one be sure that instant-propositions, as defined, are true at one instant alone? (If the quantifier-free definition of I is adopted, then the question is one of existence.) A similar problem arose for world-propositions, but only in the context of contingent propositional existence. If propositions necessarily exist then there is no harm in identifying worlds at which the same propositions are true. But the same manœuvre is not available in tense-logic, for the properties of the earlier-later may be disturbed. In other words, there is no guarantee that $-M(p \wedge Fp)$ holds for instant-propositions.

There are two familiar situations in which this failure can occur. First, there is the case of an interval in which no change occurs. Now means of distinguishing the different instants of the interval may be open to the tense-logical theorist which are not open to someone who thinks in terms of events. Suppose A occurs throughout the interval, but not before. Then $A \wedge -PA$ occurs at the first instant of the interval, $A \wedge PA \wedge -PPA$ at the second instant, and so on. But this is to suppose that there *is* a first or second instant. If the interval were open and dense, then no such description could be given. Secondly, there is the case of cycles. Suppose time is like the integers and the instants alternate between *A*- and *B*-instants thus:

A	*B*	*A*	*B*	*A*
.	.	.	.	.
—2	—1	0	1	2

Alternatively, and more plausibly, *A* and *B* could represent longish periods of history. Then, in this case too, there is no tense-logical sentence that distinguishes between the even or between the odd numbered instants.

There are various responses one can make to this difficulty. One is to deny that such situations can occur. Another is to tie individuals to times. Thus if Bill is smoking a pipe at Time 0 in the cyclic situation above then, strictly speaking, it must be another Bill smoking his pipe at Time 2. There is some plausibility in having time-bound individuals (though none, as far as I can see, in having them world-bound). However, time-bound individuals are not in keeping with the general approach of tense-logic and, in any case, they are of no help when the domain is empty at the instants to be distinguished. Yet a third response is to identify the indiscernible instants and then restructure the earlier-later relation accordingly. Thus the dead stretch above would correspond to an instant being earlier than itself ($M(p \wedge Fp)$) and the cycle would correspond to two instants being earlier than one another ($M(p \wedge Fp) \wedge M(q \wedge Fp)$). The earlier-later relation would, at least, be transitive and connected since there are tense-logical laws that guarantee these properties. Various transformations (suggested by the completeness proofs in tense-logic) could then be used to obtain linear orderings, though these orderings would be far from unique. Another response altogether is to distinguish the different instants by Now. For example, in the cyclic situation, what distinguishes instants 0 and 2 is that at 0 *A* is happening *now*. I treat this response with some misgiving. For, first, I do not accept the corresponding argument in modal logic—that things are *actually* happening does not distinguish the actual world from other possible worlds; and, second, the use of 'now' smacks too much of a reference to an instant. Perhaps one should just accept that instant-propositions have the desired properties, even though no expressible propositions have those properties.

Let us now consider a language in which both tense-logical and modal operators are used together.[1] If the Priority Thesis for time and modality is accepted, then such a language is very natural. For modal operators will apply to tensed sentences, unless sentences subject to modal qualification are specially restricted; and similarly for the application of tense-logical operators. The original language, then, will be like the earlier modal languages, but with the three intensional operators G, H and $\Box$. The classical language will now contain three sorts—for individuals, possible worlds and instants. To each n-place predicate R, there will be an $(n + 2)$ place predicate R' with extra arguments for a world and instant. The classical translation is then as follows: Rxy goes into $R'\text{xy}wt$; GB into $\forall u(t < u \supset B'(w, u))$, and similarly for H; $\Box B$ into $\forall v B'(v, t)$; and $\forall x B$ into $\forall x(E'wt \supset B'(\text{x}, w, t))$. This semantics is sometimes called *two-dimensional* since it states the truth-conditions relative to a world-instant pair.

It is perhaps worth emphasising that all modal notions are now suitably tensed. Thus the modal operators apply to tensed sentences to form tensed sentences, as in 'Possibly I will not exist'. Also, the domain of individuals is relative to the world at a time. There are not now two ontologies, one for time and the other for modality, but a single ontology for both. That only actuals or that only present individuals exist are now but two aspects of this one ontology.

In reversing the translation, the obvious strategy is to synthesise the earlier translations for time and modality. Thus world-propositions are defined by Q, instant-propositions by I, and the predication $R'\text{xy}wt$ is replaced with $L\Box(p \wedge q \supset Rxy)$, where p is a world-proposition and q is an instant-proposition. But this will not do. For within the context of our two-dimensional language, the Q-propositions are those true in one world at the *present* instant and the I-propositions are those true at one instant in the *actual* world; and these definitions have unfortunate consequences, both technical and philosophical.

[1] Such a language is briefly mentioned in *Papers on Time and Tense*, p. 133.

The technical misfortune is that the account of predication requires that $p \wedge q$ is true at a single world-instant if p is a world-proposition and q is an instant-proposition. But once we stray from the actual world or the present instant there is no guarantee that this is so on the definitions above.

The philosophical misfortune is that the definitions do not respect the necessary features of instants or the sempiternal features of worlds. The *only* necessary features of instants seem to be that: (a) they are instants, (b) they exist, and (c) they have a certain structure, i.e. that if the instant is earlier than another then it is necessarily earlier. In a more sophisticated account of temporal structure, (b) might be disputed and (c) be made conditional on the existence of the instances. However, on our perhaps over-simple account, (a)–(c) should certainly hold. Now instant-propositions, as defined above, may not possess these necessary features. (a) corresponds to an instant-proposition p being necessarily one, (b) to p being necessarily true; and it is clear that neither (a) nor (b) need hold. (c) corresponds to $L(p \supset Fq)$ being necessarily true when true. Now if Fq is part of the description of p, as when the tensed-descriptions are complete, then (c) will hold; but not otherwise.

World-propositions possess corresponding sempiternal features, though (c) does not apply. However, the *necessary* features of possible worlds are not determined by the counterparts to (a) and (b). For, first, possible worlds (i.e. world-propositions) do not necessarily exist. But, second and more importantly, possible worlds possess other necessary features. What goes on in a possible world necessarily goes on in that world; indeed, the identity of a world is determined by its content. But what goes on at an instant is purely, or largely, accidental; the identity of an instant is determined, if at all, by its position in the temporal structure.

In order to overcome these difficulties, it is necessary to revise the definitions of instant- and world-propositions. An instant-proposition p is now true of one and the same instant in *each* possible world. In modal-tense-logical terms, the definition of Ip is $M\Box(p \wedge G - p \wedge H - p)$; p is an instant-proposition if at some time it is necessarily present and necessarily not future or past. If time is not linear, then the definition could be $M\Box(p \wedge \forall q (q \supset L(p \supset q)))$. Similarly, a world-proposition is now true of one and the same world at each instant. The definition of Qp is $\Diamond L(p \wedge \forall q(q \supset \Box(p \supset q)))$, in exact analogy to the second definition of Ip. Thus instant-propositions and world-propositions are the temporal and modal cross-sections, respectively, of the two-dimensional instant-world manifold.

The reverse translation can now go as follows: $R\mathrm{xy}wt$ is replaced with $L\Box(p \wedge q \supset Rxy)$, where p and q are the world- and instant-propositions corresponding to w and t respectively; $t < u$ is replaced with $L\Box(q \supset Fr)$, where q and r are the instant-propositions corresponding to t and u; $\forall \mathrm{x}$ is replaced with $L\Box\forall x$; and $\forall w$ and $\forall t$ are replaced with quantifiers $\forall p$ and $\forall q$ over world- and instant-propositions respectively.

An alternative, but equivalent, account can be given by defining the propositional counterpart to world-instant pairs. Say that p is a *state-proposition* if it is true at exactly one world-instant pair, i.e. if $M\Diamond(p \wedge \forall q(q \supset L\Box(p \supset q)))$ holds. Thus a state-proposition corresponds to the state of the world at an instant, though the state is not a time-slice, since that can reappear in other worlds, but *everything* that is presently the case at the given instant. Say that state-propositions p and q *agree on worlds* if $\Diamond(p \wedge Mq)$ holds and that they *agree on instants* if $M(p \wedge \Diamond p)$ holds. Then instead of quantifying over instant- or world-propositions in the reverse translation, one can quantify over all state-propositions that agree on instants or on worlds with the given state-proposition. It is as if instants and worlds are identified with suitable equivalence classes of world-instant pairs.

Any world- or instant-proposition in the new sense is also such a proposition in the old sense. Therefore there will be the same problems of existence and, also, the same solutions to those problems. However, the new sense brings its own existence-problems too. An instant-proposition is true in all possible worlds at a given instant; it is necessarily true at that instant. So any sentence expressing the proposition cannot refer to any contingent feature of the world; it cannot say anything significant about what is happening at the instant. But the only content-free sentences of this sort are those to the effect that things are happening *now*. Thus even without cycles or dead time, the present approach is committed to now-sentences or the inexpressibility of instant-propositions.

The world propositions are subject to new problems over

individual existence. Suppose I attempt to express the true world-proposition, i.e. I attempt to give a description of the world that, at all times, implies all truths. Now I may need to say that at some time an individual visits Mars, and yet not be able to say which individual since there have been no past or present Martian visitors. But assume that another possible world is the same as this one but for the identity of the merely future individuals and that, in particular, the first visitor to Mars is a different individual. Then at the time of the first Martian visit, my description will no longer imply all truths since it will not imply that one individual was the visitor rather than another. Thus the world-proposition may change over time. As the world becomes more definite, then so do the propositions that describe it. But at no time may there be a proposition that describes the world, in its temporal entirety, with complete definiteness.

Under these circumstances, the earlier reduction will not work. For at a given time, a world-proposition will not pick out a single world but also all the other future 'realisations' of that world. The reduction can be repaired if the quantifiers are cumulative and the falsehood convention is adopted. Instead of determining whether a formula A is true at all instants in a given world, one deals with the instants at which the individuals involved in A exist by ordinary means and then deals with the other instants by means of the Falsehood Convention. Formally speaking, if B is a classical formula whose sole variable is **x** then the translation of $\forall w B$ is $L\, \forall p(\underline{Q}\, p \supset (\Box(p \supset Ex) \supset B^*) \wedge (-\Box(p \supset Ex) \supset B^f))$, where B^f is the result of replacing atomic sentences containing free occurrences of x in B^* by $\perp$.

Our interest in modal tensed logic has centred on the reduction problem. However, many problems in the metaphysics of time involve possibility and are naturally discussed within the context of such a language. My aim is not to discuss the nature of time as such, but let me end with one illustration of this type of application. It is that there is an intimate connection between tense-logical priority and the view that time is absolute. Time may be absolute in two senses. It may be absolute in the sense that simultaneity is well-defined without reference to any physical system (the intra-world sense). Or it may be absolute in the sense that simultaneity is well-defined without reference to a possible world, i.e. that there is simultaneity between events of different possible worlds (the inter-world sense).

Now once the tense-logical theorist combines tense and modality he is almost committed to the view that time is absolute in the second sense.[1] For in order to interpret 'Possibly I am not sitting', he must suppose that in some possible world I am not sitting *at the same time* as I am sitting in the actual world. One could interpret this sentence as saying that another possible world first diverges from the actual world in the fact that I am not sitting. But this type of interpretation would not work for more complicated possibility sentences. To take an extreme example, the tense-logical theorist may claim that in the previous cycle it is possible that *B* occurs when *A* does and *A* occurs when *B* does. For at Time 0, he may claim that it is possible that: *B* occurs, *A* occurs tomorrow, *A* occurs yesterday, and so on.

For the tense-logical theorist, the present has objective significance and he may secure cross-world simultaneity by supposing that at each time the same present runs through each possible world. In this respect, he has the edge over someone who would believe that space is absolute; for it is most implausible to suppose that there is an objective 'here' or that spatial indexicals are primitives.

[1] Consult *Papers on Time and Tense*, pp. 133–4, for a discussion of how tense-logic may be reconciled with relativity physics.

Technical Appendix

This appendix gives a firm technical footing to some of the translations. For reasons of space, only the key translations are considered.

1. *The first translations*

The translations combine an intensional account of possible worlds with a pseudo-quantifier approach to possible individuals. The propositional quantifiers are sentential and the propositions are assumed to exist necessarily.

Languages. There are three languages: the original modal language $\mathscr{L}$, the classical language $\mathscr{L}'$, and the secondary modal language $\mathscr{L}^*$. $\mathscr{L}$ is the language of quantified modal logic. It contains various predicates of specified degree, the individual variables $x_1, x_2, \ldots$, the universal quantifier $\forall$, the truth-functional connectives v and —, the modal connective $\Box$, and brackets. We assume that the predicates include the one-place predicate E of existence and the two-place predicate I of identity. Formulas are defined in the usual way.

$\mathscr{L}'$ is a classical two-sorted first-order language. It contains: two sorts of variables—$\mathrm{x}_1, \mathrm{x}_2, \ldots$ for possible individuals and $w_1, w_2, \ldots$ for possible worlds; the constant ω for the actual world; for each predicate R of degree n in $\mathscr{L}$, a predicate R' of degree $(n + 1)$ whose first n arguments are for individuals and whose last argument is for worlds; an identity predicate on individuals; the logical symbols v, — and $\forall$, and brackets. Again, formulas are defined in the usual way.

$\mathscr{L}^*$ is obtained from $\mathscr{L}$ by adding propositional variables $p_1, p_2, \ldots$ To the definition of formula are added the clauses: p_i is a formula for $i = 1, 2, \ldots$; if A is a formula then so is $\forall p_i A$.

Translations. If x (or y_i) is the variable x_j, let $\mathbf{x}$ (or $\mathbf{y}_i$) be the variable $\mathbf{x}_j$. The classical translation t takes an $\mathscr{L}$-formula A and a world-variable or constant w into an $\mathscr{L}'$-formula $t(A,w)$. t is defined by the following clauses:

(i) $t(Ry_1 \ldots y_n, w) = R'\mathbf{y}_1 \ldots \mathbf{y}_n w$
(ii) $t(-B, w) = -t(B, w)$
(iii) $t((B \vee C), w) = (t(B, w) \vee t(C, w))$
(iv) $t(\forall xB, w) = \forall \mathbf{x}(E'\mathbf{x}w \supset t(B, w))$
(v) $t(\Box B, w) = \forall w_1, t(B, w_1)$

t is defined on formulas alone by the clause:

(vi) $t(A) = t(A, \omega)$

For A in $\mathscr{L}^*$, let QA abbreviate $\Diamond(A \wedge \forall p(p \supset \Box(A \supset p))$ where p is the first propositional variable not to occur free in A. The reverse translation s takes ω-free formulas of $\mathscr{L}'$ into $\mathscr{L}^*$-formulas. Its defining clauses are:

(i) (a) $s(R'\mathbf{y}_1 \ldots \mathbf{y}_n w_i) = \Box(p_1 \supset \mathrm{R}y_1 \ldots y_n)$
(b) $s(\mathbf{y}_1 = \mathbf{y}_2) = y_1 = y_2$
(ii) $s(-B) = -s(B)$
(iii) $s((B \vee C)) = (S(B) \vee s(C))$
(iv) $s(\forall w_i B) = \forall p_i(Q\, p_i \supset s(B))$
(v) $s(\forall \mathbf{x} B) = \Box\, \forall x s(B)$

All subsequent translations will satisfy the analogues of clauses (ii)–(iii) and therefore this part of the definition will be omitted.

s is extended to formulas A containing ω in the following way. Let A^f be the result of replacing each occurrence of ω in A by the first variable w_i not to occur free in A. Then:

$$s(A) = \exists\, p_i(p_i \wedge Q\, p_i \wedge s(A^f)).$$

This translation is equivalent to treating ω as a description $\iota p_i(p_i \wedge Q\, p_i)$ with widest scope, for the uniqueness condition is automatically satisfied.

Semantics. A *model for* $\mathscr{L}$ is a quadruple $\mathfrak{M} = (W, D, w_o, \phi)$ such that:

(a) W (worlds) and D (possible individuals) are non-empty sets, w_o (actual world) ε W, and $\phi(R, w) \subseteq D^n$ for each n-place predicate R and $w \, \varepsilon \, W$;
(b) $\phi(I, w) = \{\langle a, a\rangle : a \, \varepsilon \, D\}$ for each $w \, \varepsilon \, W$;
(c) $(\forall \, a \, \varepsilon \, D)(\exists \, w \, \varepsilon \, W)(a \, \varepsilon \, \phi(E, w))$.

Let $D_w = \{a : a \, \varepsilon \, \phi(E, w)\}$ and call θ an assignment if it is a function from the set of individual variables into D. Then the definition of satisfaction contains the following clauses:

(i) $\mathfrak{M} \models_\theta R y_1 \ldots y_n$ iff $\langle \theta(y_1), \ldots, \theta(y_n)\rangle \, \varepsilon \, \phi(R, w_o)$
(ii) $\mathfrak{M} \models_\theta \Box B$ iff $\mathfrak{M}' \models_\theta B$ for all models $\mathfrak{M}'$ differing from $\mathfrak{M}$ in the third component only
(iii) $\mathfrak{M} \models_\theta \forall x_i B$ iff $\mathfrak{M} \models_{\theta_1} B$ for all θ' such that $\theta'(x_i) \, \varepsilon \, D w_o$ and $\theta'(x_j) = \theta(x_j)$ for all $j \neq i$.

The clauses for v and $-$ are standard.

Given an $\mathscr{L}$-model $\mathfrak{M} = (W, D, w_o, \phi)$, let ϕ' be the function ψ such that $\psi(R') = \{\langle a_1, \ldots, a_n, w\rangle : \langle a_1, \ldots, a_n\rangle \, \varepsilon \, \phi(R, w)\}$ for each predicate R' in $\mathscr{L}'$ and let $\mathfrak{M}' = (W, D, w_o, \phi')$. Then a *model for* $\mathscr{L}'$ is a model $\mathscr{C} = \mathfrak{M}'$ where $\mathfrak{M}$ is a model for $\mathscr{L}$. In $\mathfrak{M}'$ W is now regarded as the domain of the world-variables, D as the domain of the individual-variables, w_o as the denotation of ω, and ψ as the valuation. The truth-definition then proceeds along classical lines.

A *model for* $\mathscr{L}^*$ is a quintuple $\mathscr{C} = (W, D, P, w_o, \phi)$, where $\mathfrak{M} = (W, D, w_o, \phi)$ is an $\mathscr{L}$-model and P (propositions) is a subset of $\mathfrak{P}(W)$. Such a model is *complete* if $P = \mathfrak{P}(W)$. θ, α is an *assignment for* $\mathscr{C}$ if α is an assignment for $\mathfrak{M}$, as above, and α is a function from the set of propositional variables $p_1, p_2, \ldots$ into P. In the definition of satisfaction, the clause for $\forall \, p_i$ is:

$\mathscr{C} \models_{\theta,\alpha} \forall p_i B$ iff $\mathscr{C} \models_{\theta,\alpha} B$ for all α' such that $\alpha'(p_i) \, \varepsilon \, P$ and $\alpha'(p_j) = \alpha(p_j)$ for all $j \neq i$.

The other clauses are as before.

Correctness of translations. We state results which show that the translations are correct and compatible with one another.

Given an $\mathscr{L}$-model $\mathfrak{M} = (W, D, w_0, \phi)$, let $\mathfrak{M}^*$ be the $\mathscr{L}^*$-model $(W, D, \mathfrak{P}(W), w_0, \phi)$.

Theorem 1. (a) For any $\mathscr{L}$-sentence A and $\mathscr{L}$-model $\mathfrak{M}$, $\mathfrak{M} \models A$ iff $\mathfrak{M}' \models t(A)$.
(b) For any $\mathscr{L}'$-sentence A and $\mathscr{L}'$-model $\mathfrak{M}'$, $\mathfrak{M}' \models A$ iff $\mathfrak{M}^* \models s(A)$.
(c) For any $\mathscr{L}$-sentence A and $\mathscr{L}$-model $\mathfrak{M}$, $\mathfrak{M} \models A$ iff $\mathfrak{M}^* \models s(t(A))$.

Similar results hold for formulas; indeed, these are required for an inductive proof of the above theorem.

Say that a sentence A of $\mathscr{L}$ ($\mathscr{L}'$, $\mathscr{L}^*$) is *valid* if it is true in all $\mathscr{L}$-models (respectively: $\mathscr{L}'$-models, complete $\mathscr{L}^*$-models). Then the theorem has the following consequences:

Corollary 2. For sentences A from the appropriate language:
(a) A is $\mathscr{L}$-valid iff $t(A)$ is $\mathscr{L}'$-valid
(b) A is $\mathscr{L}'$-valid iff $s(A)$ is $\mathscr{L}^*$-valid
(c) A is $\mathscr{L}$-valid iff $s(t(A))$ is $\mathscr{L}^*$-valid
(d) $A \equiv s(t(A))$ is $\mathscr{L}^*$-valid.

The above results also hold if complete $\mathscr{L}^*$-models are replaced with *atomic* models, i.e. ones in which P is an atomic Boolean algebra.

Axiomatisation. The sets of $\mathscr{L}$-valid and $\mathscr{L}'$-valid formulas can be axiomatised. An axiomatisation of the $\mathscr{L}$-valid formulas can be found in [2]. An axiomatisation of the $\mathscr{L}'$-valid formulas can be obtained by adding the following axioms to the appropriate two-sorted predicate calculus:

$$I'\mathrm{x}_1\mathrm{x}_2 w_1 \equiv \mathrm{x}_1 = \mathrm{x}_2$$
$$\forall \mathrm{x}_1 \exists w_1 E'\mathrm{x}_1 w$$

These axioms express the conditions (b) and (c) on an $\mathscr{L}$-model; and so completeness follows from the completeness result for the predicate calculus.

Corollary 2 now implies a syntactic counterpart to (a):

Corollary 3. A is a theorem of the system for $\mathscr{L}$ iff $t(A)$ is a theorem of the system for $\mathscr{L}'$.

This result also has, of course, a direct syntactic proof.

The set of $\mathscr{L}'$-valid formulas is not axiomatisable. However, it may be determined which axioms are required for the syntactic counterparts of corollary 2 (b)–(d) to hold. First, add to the system for $\mathscr{L}$ the standard quantificational principles for $\forall p_i$, viz. $\forall p_i A \supset Ap_i/B$, B free for p_i in A, and $\forall p_i(B \supset C) \supset (B \supset \forall p_i C)$, p_i not free in B. Then add the following non-logical axioms:

Existence of world-propositions $\Box \exists p(p \wedge Q\, p)$
Barcan formula $\Box \forall p A \equiv \forall p \Box A$
ω-completeness $\Box \forall p(Q\, p \wedge \Box(p \supset \exists x B) \supset \Diamond \exists x \Box(p \supset B))$

The results then follow:

Theorem 4. For sentences A from the appropriate language:
(a) A is an $\mathscr{L}'$-theorem iff $s(A)$ is an $\mathscr{L}^*$-theorem
(b) A is an $\mathscr{L}$-theorem iff $s(t(A))$ is an $\mathscr{L}^*$-theorem
(c) $A \equiv s(t(A))$ is an $\mathscr{L}^*$-theorem.

Dropping individual variables. Further results can be obtained if individual variables are dropped and consequent changes are made in the languages, models and translations. $\mathscr{L}$ is now the language of modal sentential logic, $\mathscr{L}'$ the language of the monadic predicate calculus (with world-variables only), and $\mathscr{L}^*$ the language of $\mathscr{L}$ with sentence variables.

First, the logic for $\mathscr{L}^*$ is now decidable and has a particularly simple axiomatisation.[1]

Second, a reverse translation into $\mathscr{L}$ is now possible. Say that an $\mathscr{L}'$-formula is *uniform* if it contains no world-variables other than w_1. Then note that $t(A)$ is always a uniform formula. In case individual variables are dropped, a translation of uniform $\mathscr{L}'$-formulas into $\mathscr{L}$ can be obtained by replacing $\forall w_1$ with $\Box$ and

[1] See my 'Propositional Quantifiers in Modal Logic', *Theoria*, 36 (1970), pp. 336–46, and the references therein.

$P'\omega$ with P (but after $P'\omega$ has been moved outside the scope of a $\forall w$ quantifier).[1] Since each formula of the monadic predicate calculus is equivalent to a uniform one, this gives a full reverse translation. However, if individual variables are retained, then the translation of uniform $\mathscr{L}'$-formulas into $\mathscr{L}$ is no longer possible. For example, $\forall w_1 \exists x R' x w$ has no translation in $\mathscr{L}'$.

Translations without necessarily existing propositions
These translations extend the earlier ones to the case in which propositions do not necessarily exist.

Languages. $\mathscr{L}$ and $\mathscr{L}'$ are as before. The secondary modal language, now dubbed $\mathscr{L}^\circ$, is obtained from $\mathscr{L}$ by adding: propositional variables $\rho_1, \rho_2, \ldots$; and the predicates T' for existence and $\rightarrow'$ for strict implication. The variables ρ_i are now nominal and so to the definition of formula in $\mathscr{L}$ is added the clause: $T'\rho$, $\rho \rightarrow' \sigma$, $E\rho$ and $\rho = \sigma$ are formulas when ρ and σ are propositional variables. The proposition—that operator ξ is not required for the reduction. However, if it were added to the language, one would need to distinguish between terms and formulas. Propositional variables are terms, and so is ξA for A a formula; $T'm$, $m \rightarrow' n$, etc. are formulas for arbitrary terms m and n.

Translations. Let $Q\ \rho_i$ now abbreviate $\Diamond(T'\rho_i \wedge \forall \rho_{i+1}(T'\rho_{i+1} \supset (\rho_i \rightarrow' \rho_{i+1})))$. The reverse translation r from ω-free formulas of $\mathscr{L}'$ into $\mathscr{L}^\circ$ is then defined by:

(i) (a) $\mathrm{r}(R'\mathrm{y}_1 \ldots \mathrm{y}_n w_i) = \Box\, (T'\rho_i \supset Ry_1 \ldots y_n)$
    (b) $\mathrm{r}(\mathrm{y}_1 = \mathrm{y}_2) = y_1 = y_2$
(ii) $\mathrm{r}(\forall w_i B) = \Box\, \forall \rho_i(Q\ \rho_i \supset \mathrm{r}(B))$
(v) $\mathrm{r}(\forall \mathrm{x} B) = \Box\, \forall x \mathrm{r}(B)$.

As before, let A^f be the result of replacing each occurrence of ω in A by the first world variable w_i not to occur free in A. Then r is extended to all $\mathscr{L}'$-formulas A by the clause:

$$\mathrm{r}(A) = \exists \rho_i(T'\rho_i \wedge Q\ \rho_i \wedge \mathrm{r}(A^f)).$$

Semantics. A *model for* $\mathscr{L}^\circ$ is a sextuple $\mathscr{C} = (W, D, P, w_o, \phi, x)$,

[1]See the last sections of Chapter 2 above.

where (W, D, P, w_o, ϕ) is an $\mathscr{L}^*$-model and X is a function from W into $\mathfrak{P}(P)$ subject to the condition that $(\forall X \in P)(\exists w \in W)(X \in \mathrm{X}(w))$. Intuitively, $\mathrm{X}(w)$ is the domain of propositions that exist in world w. The model $\mathscr{C}$ is *complete* if each $\mathrm{X}(w)$ is a complete Boolean algebra and *atomic* if $\{w\} \in \mathrm{X}(w)$ for each w. Note that a complete model need not be atomic.

An *assignment* θ, α *for* $\mathscr{C}$ is defined as before, but with $\rho_1, \rho_2, \ldots$ replacing $p_1, p_2, \ldots$ in the domain of α. The note-worthy clauses in the truth-definition are:

(i) (a) $\mathscr{C} \vDash_{\theta,\alpha} T'\rho_i$ iff $\alpha(\rho_i) \in \mathrm{X}(w_o)$ and $w_o \in \alpha(\rho_i)$
(b) $\mathscr{C} \vDash_{\theta,\alpha} \rho_i \rightarrow' \rho_j$ iff $\alpha(\rho_i), \alpha(\rho_j) \in \mathrm{X}(w_o)$ and $\alpha(\rho_i) \subseteq \alpha(\rho_j)$
(c) $\mathscr{C} \vDash_{\theta,\alpha} E\rho_i$ iff $\alpha(\rho_i) \in \mathrm{X}(w_o)$
(d) $\mathscr{C} \vDash_{\theta,\alpha} \rho_i = \rho_j$ iff $\alpha(\rho_i) = \alpha(\rho_j)$
(ii) $\mathscr{C} \vDash_{\theta,\alpha} \forall \rho_i B$ iff $\mathscr{C} \vDash_{\theta,\alpha'} B$ for all α' such that $\alpha'(\rho_i) \in \mathrm{X}(w_o)$ and $\alpha'(\rho_j) = \alpha(\rho_j)$ for all $j \neq i$.

If the language $\mathscr{L}^\circ$ contained ξ, then the denotation $d(m)$ of each term m would need to be determined: $d(\rho_i) = \mathrm{X}(\rho_i)$; and $d(\xi A) = \{w \in W: \mathscr{C} \vDash_{\theta,\alpha} A\}$. Clause (i) could then be formulated with denotations of arbitrary terms replacing the assignment function α.

Correctness. Say then an $\mathscr{L}^\circ$-model $\mathscr{C}$ *extends* an $\mathscr{L}'$-model $\mathscr{M}$ if $\mathscr{M}$ is $\mathscr{C}$ without its last component. Then in analogy to theorem 1 (b)–(c), we have:

Theorem 5. Suppose $\mathscr{M}$ is an $\mathscr{L}$-model and $\mathscr{C}$ an atomic $\mathscr{L}^\circ$-model that extends $\mathscr{M}$. Then:

(a) $\mathscr{M}' \vDash A$ iff $\mathscr{C} \vDash \mathrm{r}(A)$ for any $\mathscr{L}'$-sentence A
(b) $\mathscr{M} \vDash A$ iff $\mathscr{C} \vDash \mathrm{r}(t(A))$ for any $\mathscr{L}$-sentence A.

It would also be possible to establish analogues of Corollaries 2 and 3 and Theorem 4. However, I shall not go into details.

Index of Symbols

Index of Names

General Index